D0747301

Special relativity

A. P. French

PROFESSOR OF PHYSICS, THE MASSACHUSETTS INSTITUTE OF TECHNOLOGY

Special relativity

THE M.I.T. INTRODUCTORY PHYSICS SERIES

W · W · NORTON & COMPANY

New York · London

W. W. Norton & Company, Inc., 500 Fifth Avenue, New York, N.Y. 10110
W. W. Norton & Company Ltd., 10 Coptic Street, London, WC1A 1PU

Library of Congress Catalog Card No. 68-12180

ISBN 0-393-09793-5

printed in the united states of america

1 2 3 4 5 6 7 8 9 0

Contents

6 Relativistic dynamics—collisions and conservation laws 167

7 More about relativistic dynamics 205

8 Relativity and electricity 229

Preface

THE WORK of the Education Research Center at M.I.T. (formerly the Science Teaching Center) is concerned with curriculum improvement, with the process of instruction and aids thereto, and with the learning process itself, primarily with respect to students at the college or university undergraduate level. The Center was established by M.I.T. in 1960, with the late Professor Francis L. Friedman as its Director. Since 1961 the Center has been supported mainly by the National Science Foundation; generous support has also been received from the Kettering Foundation, the Shell Companies Foundation, the Victoria Foundation, the W. T. Grant Foundation, and the Bing Foundation.

The M.I.T. Introductory Physics Series, a direct outgrowth of the Center's work, is designed to be a set of short books which, taken collectively, span the main areas of basic physics. The series seeks to emphasize the interaction of experiment and intuition in generating physical theories. The books in the series are intended to provide a variety of possible bases for introductory courses, ranging from those which chiefly emphasize classical physics to those which embody a considerable amount of atomic and quantum physics. The various volumes are intended to be compatible in level and style of treatment but are not conceived as a tightly knit package; on the contrary, each book is designed to be reasonably self-contained and usable as an individual component in many different course structures.

The present volume is written as an introduction to special relativity for students who have a modest background in Newtonian mechanics and an acquaintance with the rudiments of optics and electricity. The approach is traditional (for this particular level) in that it does not rest heavily on electromagnetic theory but concentrates on the problems of kinematics and dynamics. The last chapter, however, deals with some of the insights that relativity can provide with regard to the relationship between electricity and magnetism. The main substance of this book has been used successfully with both first- and second-year students at M.I.T. as part of a general introductory physics course; the extent and coverage are, however, such that the book may also be found suitable as a self-contained introduction to relativity for more advanced students.

This book, like the others in the series, owes much to the thoughts, criticisms, and suggestions of many different people, both students and instructors. In the latter category, the detailed comments of Prof. M. W. Friedlander (Washington University), Prof. A. W. K. Metzner (San Diego State College), and Prof. Rainer Weiss (M.I.T.) have been particularly helpful.

A special acknowledgment is due to Prof. Jack R. Tessman (Tufts University), who was deeply involved with our earliest work on the introductory physics series and has contributed in an especially important way to this relativity text. With the present author, he taught the first trial version of the material at M.I.T. during 1963–1964. The subsequent writing and rewriting was discussed with him in detail and embodies many of his suggestions. In particular, the final chapter, on relativity and electricity, is based largely on a much more far-reaching analysis by Prof. Tessman, in which the main results of electromagnetism, including the acceleration fields, are developed [see *Am. J. Phys.*, **34,** 1048–1055 (1966), and *Am. J. Phys.*, **35,** 523–527 (1967)].

Thanks are also due to Prof. M. K. Smith and Dr. James A. Ross for valuable assistance in the preparation of this volume.

A. P. FRENCH

Cambridge, Massachusetts
February 1968

Special relativity

In experimental philosophy we are to look upon propositions obtained by general induction from phenomena as accurately or very nearly true . . . till such time as other phenomena occur, by which they may either be made more accurate, or liable to exceptions.

SIR ISAAC NEWTON, *Principia* (1686)

The relativity theory arose from necessity, from serious and deep contradictions in the old theory from which there seemed no escape. The strength of the new theory lies in the consistency and simplicity with which it solves all these difficulties, using only a few very convincing assumptions . . . The old mechanics is valid for small velocities and forms the limiting case of the new one.

A. EINSTEIN AND L. INFELD,
The Evolution of Physics (1938)

1

Departures from Newtonian dynamics

WHAT IS IT that you first think of when you see or hear the word *relativity*? Very likely there will come to your mind the name of Albert Einstein, or the equation $E = mc^2$, or a vision of space travelers returning youthful from trips of many years' duration. This is a well-deserved tribute to the enormous intellectual impact—still effective, more than 60 years after the event—of what Einstein called his *special theory* of relativity. And the development of this theory by Einstein and others in the years around 1900 is rightly regarded as one of the greatest strides ever made in our way of describing and interpreting the physical world. Yet the basic concept of relativity is as old as the mechanics of Galileo and Newton. It is, crudely speaking, just the assertion that the laws of physics appear the same in many different reference frames. What, then, did Einstein do to make his name almost synonymous with the title of this book? The answer is that he led us to apply the notions of relativity to *all* our physical experience and not merely to a restricted range of phenomena. In particular, he asserted that processes involving very rapid motions—specifically, motions at speeds of the order of the speed of light—are not to be placed in a separate category. But the unification that he proposed brought with it some remarkable implications. There were consequences that seemed opposed to our intuitions and our common sense, in a way that classical theories were not—the increase of inertia with speed, for example, or the so-called *twin paradox*. It was such things as this that made Einstein's formulation of relativity so striking and which conferred on it a glamour and a popular interest probably never equaled in the whole history of physics.

We have said that the idea of relativity existed before Einstein and was embodied in Newton's mechanics. But it came to be recognized, about 200 years after Newton, that certain observed effects—quite small and subtle ones, for the most part—could simply not be accounted for if one tried to hold on to all the basic features of Newtonian mechanics. Historically the recalcitrant facts, demanding a revision of ideas, made their appearance in electromagnetic phenomena, especially in the propagation of light. It quickly became clear, however, primarily through Einstein's own work, that all of dynamics, and not merely the specialized field known as electrodynamics, was affected.

It was typical of Einstein, and a sign of his greatness, that he drew conclusions of the most profound and far-reaching kind from a bare minimum of data. Lesser men often attempt the same thing, of course, but differ from the Einsteins of this world in that their grand conclusions or generalizations are usually false. In essence, Einstein constructed the special theory of relativity out of a single proposition, that in every observation of the passage of light from one point to another through empty space the time taken is simply the relative separation of the points divided by a universal velocity c; it depends in no way on any velocity that one's laboratory may appear to have through space. The development of relativity from this result is not difficult (once Einstein has shown the way) and is logically clear and compelling, and we shall present it in due course. It is a development that begins with optics, proceeds to a revised kinematics, and shows us how we must rewrite the dynamics of particles. But today we can appeal to an immense amount of direct evidence concerning the dynamics of particles traveling at extremely high speeds. This evidence makes it clear from the outset that we must look for a modification of the Newtonian scheme if we are to have an acceptable dynamical description of familiar particles, such as electrons, at all speeds. And in this beginning chapter we shall proceed as quickly as possible to develop some of this revised dynamics. It is only a preview, in a way, and it is admittedly short on rigor. But there may be some interest and value in seeing how a few of the key results can be at least suggested without recourse to most of the formalism of relativity theory.

Our very first task, however, will be to give a reminder of what it is that we are going to modify, for without this the relation between the old dynamics and the new—the amount they have in common, as well as their divergences—cannot be fully appreciated and understood.

NEWTON

Newton's mechanics concerns itself with the motions of particles under the action of forces. A particle is regarded as a material point; its motion is described by the position of that point in space as a function of time. It is assumed that the separate concepts of space and time are well understood, even though they defy adequate definition. Newton believed in an absolute space, but he also recognized that one cannot chart the motion of a body through this space. Instead, we define the position of one body with respect to another: "And so," as he wrote in the *Principia*, "instead of absolute places and motions, we use relative ones."[1]

But despite the relativity of position and velocity, we do encounter an apparently absolute or fundamental quantity in the acceleration. And Newtonian dynamics seizes upon the acceleration, **a**, and relates this to the force, **F**, supplied by a particle's environment. This is an immensely fruitful procedure, because it is found that a single, constant property of the particle—its inertial mass, m—serves to connect the acceleration of the particle with the force, through $\mathbf{F} = m\mathbf{a}$. If the value of **F** is given by an explicit law of force—as in the case of universal gravitation—classical mechanics acquires the status of a physical theory, and Newton's law becomes much more than a definition of **F** in terms of m and **a**.[2]

[1] See Sir I. Newton, *Mathematical Principles of Natural Philosophy and His System of the World* (*Principia*), translated by A. Motte, revised by F. Cajori, Univ. California Press, Berkeley, 1962.

[2] See A. Einstein, "Physics and Reality," *J. Franklin Inst.*, **221**, 349–382 (1936); reprinted in Einstein's *Ideas and Opinions*, Crown, New York, 1954. See also N. Austern, *Am. J. Phys.*, **29**, 617 (1961).

Even if the law of force is not explicitly given or known, we still have one of the key statements of Newtonian mechanics—the conservation of linear momentum. Taking the inertial mass as a constant property of a body, one can verify (and this was one of the experimental foundation stones of mechanics) that the sum of the momenta $m\mathbf{v}$ for two or more interacting bodies is a constant, provided the effect of any forces of external origin can be ignored. If, for convenience, the momentum is denoted by the single vector $\mathbf{p}$, we know that in classical mechanics the ratio p/v for a given particle represents a single invariable quantity.

Finally, going beyond the strict confines of the Newtonian scheme, we have the principle of conservation of energy. Given any particular law of force, we find that the work done on any particle is reflected in a corresponding change of its kinetic energy:

$$\int_1^2 \mathbf{F} \cdot d\mathbf{r} = \tfrac{1}{2}m(v_2^2 - v_1^2) \qquad (1\text{–}1)$$

Furthermore, energy that has been conferred on a particle in this way may be recovered in a different form, as, for example, by bringing the particle to rest in a medium, with the liberation of thermal energy (i.e., heat). Our faith in the conservation of energy is so great (because of a vast body of internally consistent evidence) that we would not hesitate, in the example just mentioned, to regard the measurement of the heating as being tantamount to a measurement of the particle's kinetic energy prior to impact—provided, of course, we had reason to ignore the possibility of significant energy losses through radiation, sound, mechanical deformation, and so on.

These concepts, then, of space and time, of force, acceleration and inertial mass, of momentum and energy, comprise the foundations of classical mechanics. Now let us look at some of the cracks that have become manifest in that structure after about 200 years of apparently flawless existence. Most of them (but not all) appear in connection with the motion of particles at extremely high speeds.

"THE ULTIMATE SPEED"

According to the equations of Newtonian mechanics, there is in principle no upper limit to the velocity that may be given to an object. Imagine, for example, that a body is acted on continually by a constant force equal in magnitude to the force of gravity at the earth's surface. Its acceleration would always have the value 9.8 m/sec^2. After 1 year, starting from rest, its speed would be about 3×10^8 m/sec (i.e., equal to the speed of light in vacuum); after 2 years it would be 6×10^8 m/sec, and so on. (Take a moment to verify these numbers for yourself.) If the object were small, one could readily envisage a force that was many times larger than mg, bringing about these increases of velocity much more quickly—perhaps in a matter of minutes or seconds. Even if the force were not constant, one would be able to calculate the total amount of work, equal to the gain of kinetic energy K, required to cause a body of mass m to travel with any specified speed v: $v = (2K/m)^{1/2}$. But when the attempt is made to accelerate particles to speeds as large as those mentioned above, a drastic departure from the predictions of Newtonian mechanics is observed. We shall take this phenomenon as our first clear example of the fact that classical mechanics is not adequate for all dynamical situations.

Because of its very small mass in relation to its charge, the electron is readily accelerated to very high speeds—higher by many orders of magnitude than anything in our normal experience. Thus, for example, an electron traveling from cathode to anode of a vacuum tube, with a mere 100 volts between these electrodes, would (if it started from rest) arrive at the anode with a speed of about 6000 km/sec (and its acceleration, if the electrodes were spaced by a few millimeters, would be about $10^{15}g$). Even under these conditions the Newtonian mechanics meets the situation quite well. But if the acceleration is through millions of volts, instead of hundreds, the need for a revised dynamics becomes glaringly obvious. This has been demonstrated in a filmed experiment that explores the relation between speed and kinetic energy for electrons of kinetic energies up to 15 MeV.[1]

The experimental arrangement is shown schematically in Fig. 1–1. The experiment consists in making direct measurements of the time of flight for electrons traveling through a linear accelerator (*linac*, for short). The electrons can be given energies up to 1.5 MeV by the purely electrostatic action of a Van de Graaff generator that acts as an injector for the linac; they then enter the series of drift tubes of the linear accelerator proper and are timed over the flight path AB. Higher energies (up to about

[1]Film, *The Ultimate Speed*, by W. Bertozzi, Education Development Center, Newton, Mass., 1962. For a full description, see W. Bertozzi, *Am. J. Phys.*, **32,** 551–555 (1964).

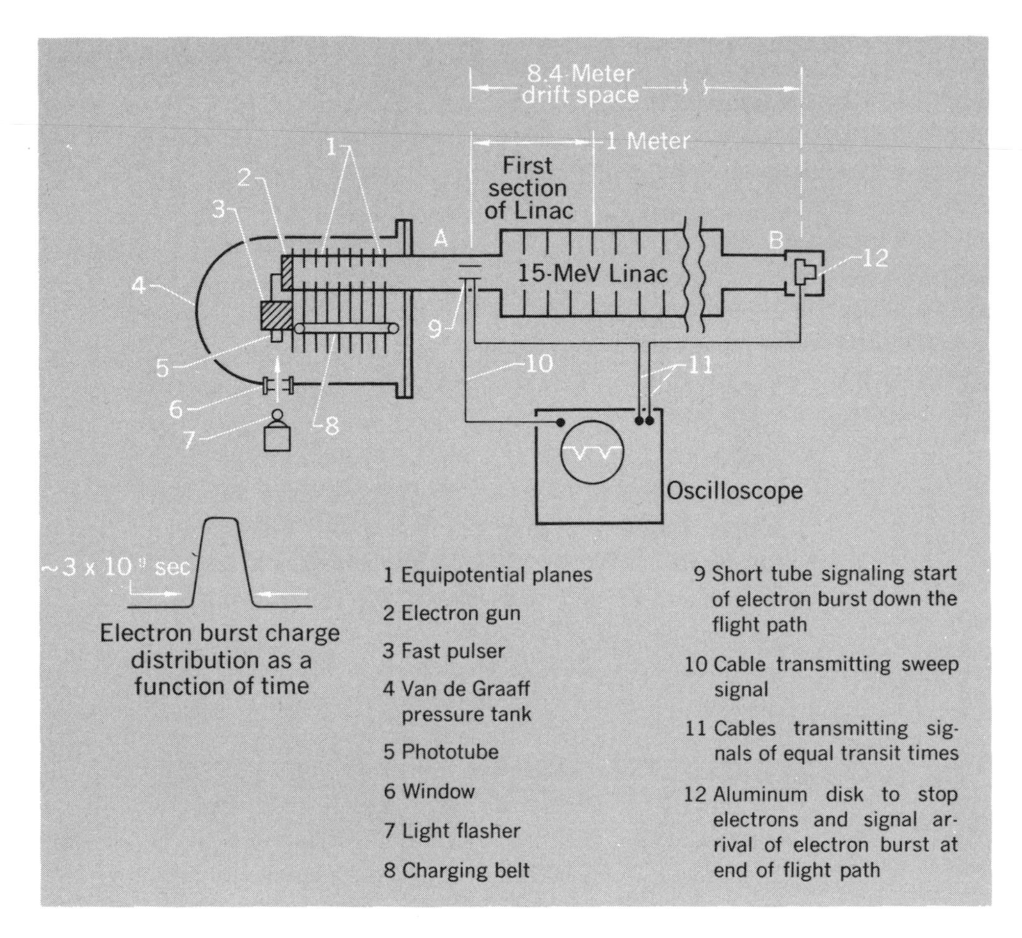

Fig. 1–1 Schematic diagram of apparatus to measure time of flight of energetic electrons. (*The "ultimate speed" experiment, by W. Bertozzi.*)

15 MeV) can be obtained by operating the radiofrequency system of the linear accelerator; in the main part of the film use is made of this for one observation only to give electrons an energy of 4.5 MeV by operating just the first section of the linac (the section immediately following the point *A*). Even in this latter situation, however, the electrons travel the whole distance *AB* with almost constant speed, as we shall see in a moment.

The electrons are released in short bursts (of about 3×10^{-9} sec duration) from the electron-gun system in the negative high-voltage terminal of the Van de Graaff accelerator. Insulated electrodes at *A* and *B* pick up electric signals as the burst passes by. These impulses are carried to a cathode-ray

TABLE 1–1

Kinetic energy *K, MeV*	*Flight time* t, $\times 10^{-8}$ *sec*	*Electron speed* v, $\times 10^8$ *m/sec*	v^2, $\times 10^{16}$ m^2/sec^2
0.5	3.23	2.60	6.8
1.0	3.08	2.73	7.5
1.5	2.92	2.88	8.3
4.5	2.84	2.96	8.8
15	2.80	3.00	9.0

oscilloscope by cables that are made equal in length. The electric signals then take equal times to reach the oscilloscope from the electrodes, so that the two pulses displayed on the oscilloscope (Fig. 1–2) provide a true measure of the time taken by the electron burst to travel from A to B. In Fig. 1–2, for example, this time (t) is about 3.3×10^{-8} sec (one main division of the horizontal scale $\approx 10^{-8}$ sec). The flight path (l) between A and B is measured to be 8.4 m. Thus from the basic definition of speed v we have

$$v = \frac{l}{t} \approx \frac{8.4}{3.3 \times 10^{-8}} \approx 2.5 \times 10^8 \text{ m/sec}$$

This measurement was for electrons accelerated through 0.5 MV (500,000 volts) by the Van de Graaff machine.

In Table 1–1 we summarize the results of the complete experiment. The most cursory inspection of these results shows that they are not at all what one would have if Newtonian mechanics were applicable. Over-all, the kinetic energy is raised by a factor of 30, so one might have looked for a factor of 5.5 in the speed (since $v \sim K^{1/2}$ according to classical mechanics). Instead, there is an increase of only about 15%. The increase of v between 1.5 and 4.5 MeV is barely detectable within the accuracy of the experiment. One might therefore question whether the

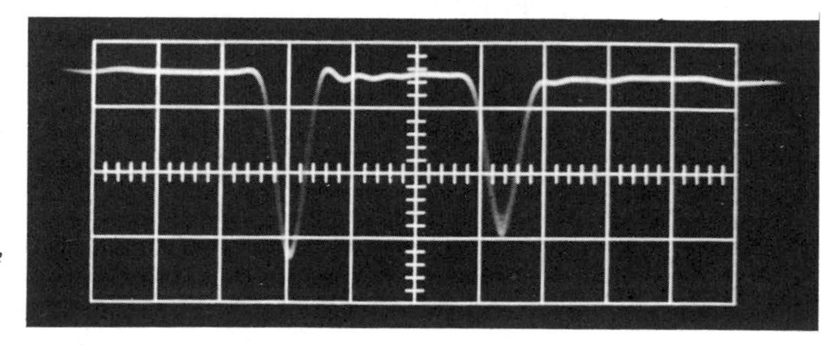

Fig. 1–2 Oscilloscope trace showing pulses due to a burst of electrons of 0.5 MeV at the beginning and end of an 8.4-m flight path. (Reproduced from the film, "The Ultimate Speed.")

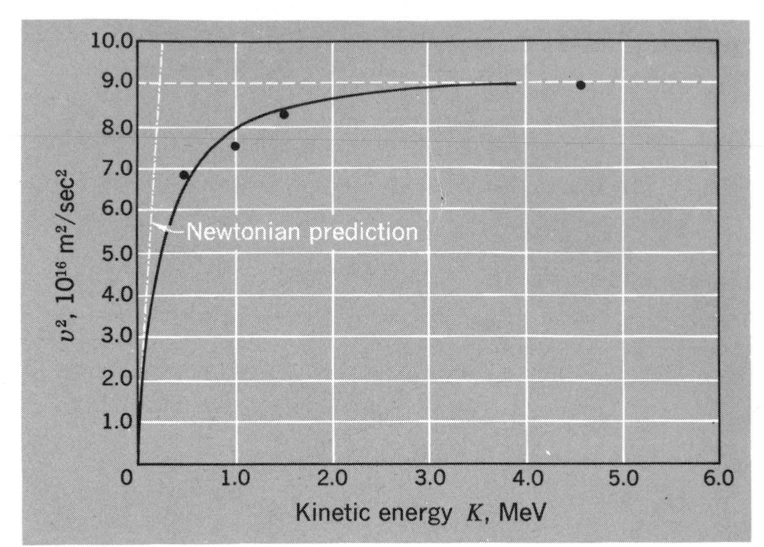

Fig. 1–3 Results of the "ultimate speed" experiment—v^2 as a function of kinetic energy for electrons—showing asymptotic approach to $v = c$.

electrons are in fact being given the energy calculated from the value of qV (charge times accelerating voltage), which successfully describes the kinetic energy gained when V is only of the order of 100 volts. In the film this question is answered by making a direct calorimetric measurement of the energy of the electrons at the point B. There is no doubt about it; the energy is there.

In Fig. 1–3 we show a comparison of the experimental results with the classical predictions. It is a graph of v^2 against K. Classically, we should have

$$v^2 = \frac{2K}{m} \tag{1–2a}$$

Numerically, this gives us

$$v^2 \ (\text{m}^2/\text{sec}^2) = 3.5 \times 10^{17} K \ (\text{MeV}) \tag{1–2b}$$

We know that this works very well for electron energies of about 1 keV or less, but we see that even at the lowest energy of the linac experiment (0.5 MeV) the value of v^2 predicted by Eq. (1–2) is too high by a factor of about 2. For higher energies the discrepancy becomes even more serious. Rather than increasing in proportion to K, the values of v^2 show all the signs of asymptotically approaching a limit, especially when one recalls the measurement at 15 MeV, not shown on the graph in Fig. 1–3. The value of v corresponding to this asymptote is 3.0×10^8 m/sec.

These results are consistent with the proposition that (to quote the last sentence of the film): "There is a speed limit for any object, and this limit is the speed of light."

The above result is very remarkable indeed. Why should one not be able to give a particle an arbitrarily high speed, if one is able to give it as much energy as one wishes? To appreciate in another way how startling this result is, suppose that electrons are being continuously accelerated in a long evacuated tube, reaching a final energy of several MeV. After the first 0.5 MeV of acceleration, the electrons have a speed of about 2.6×10^8 m/sec (cf. Table 1–1), i.e., about 85% of the speed of light. Now imagine oneself in a frame of reference moving at this speed in the same direction. In this frame the electrons at this stage of their acceleration appear to be at rest. Granted the possibility of continued acceleration, one can readily conceive of the electrons picking up energy and speed until they have the equivalent of 0.5 MeV of kinetic energy and a speed of 2.6×10^8 m/sec with respect to this new frame. But should not this mean that, as observed in the laboratory, the electrons at this stage have a speed of 5.2×10^8 m/sec, or about 1.7 times c? That is what our ordinary rules of velocity addition would suggest, but it does not happen, as the ultimate-speed experiment shows. The behavior of the electrons, as studied via measurements made throughout in the laboratory frame, demands a fundamental revision of the rules for combining velocities, i.e., the rules by which a given motion is described from the standpoint of different reference frames. We must find a new version of kinematics to deal with this. You may wish to pursue this question immediately. If so, proceed at once to Chapter 2. In the remainder of this present chapter, however, we shall explore further some of the dynamical questions raised by the ultimate-speed experiment. In particular, since the limiting speed of electrons is equal to the speed of light, we shall take a close look at the dynamics of what one may call the particles of light, i.e., photons.

PHOTONS

The speed of light, c, has long been recognized as one of the fundamental constants of nature. But it acquires a new interest when we have a photon picture of radiation. Is it really true that these photons—particles characterized by the radiation

TABLE 1-2: SPEED OF PHOTONS

Frequency, sec^{-1}	*Photon energy, eV*	*Wavelength, m*	*Speed (with error), $\times 10^8 m/sec$*
4.7×10^7	1.9×10^{-7}	6.4	2.9978 ± 0.0003
1.7×10^8	7.0×10^{-7}	1.8	2.99795 ± 0.00003
3.0×10^8	1.2×10^{-6}	1.0	2.99792 ± 0.00002
3.0×10^9	1.2×10^{-5}	1.0×10^{-1}	2.99792 ± 0.00009
2.4×10^{10}	1.0×10^{-4}	1.2×10^{-2}	2.997928 ± 0.000003
7.2×10^{10}	3.0×10^{-4}	4.2×10^{-3}	2.997925 ± 0.000001
5.4×10^{14}	2.2	5.6×10^{-7}	2.997931 ± 0.000003
1.2×10^{20}	5.1×10^5	2.5×10^{-12}	2.983 ± 0.015
4.1×10^{22}	1.7×10^8	7.3×10^{-15}	2.97 ± 0.03

frequency ν—all have exactly the same speed c, although their energies $h\nu$ may vary over a colossal range? The answer, as far as all our experience goes, is yes. Table 1–2 collects some results whose total span represents almost a factor of 10^{15} in the photon energy.[1] It may be seen that the accuracies of the results for different photon energies differ widely. The most accurate determinations are for visible light and for microwaves of about 1 cm wavelength; the photon energies differ by a factor of 10^4 but the speeds are the same to 1 part in 10^6.[2] And from the first and the last entries, we see that television transmission photons of about 10^{-7} eV and gamma rays of about 100 MeV have the same speed to an accuracy of 1%, despite the energy factor of 10^{15}. This is clearly a result of the first importance, and stands in contrast to the systematic increase of speed with energy for such particles as electrons. We may note, however, that the results of the ultimate-speed experiment make the contrast less abrupt, for it suggests (and a great body of other experience confirms) that an electron with a kinetic energy of a few MeV has a speed within 1% of the speed of light, c, and that no in-

[1]For references, and excellent accounts of experiments, see J. F. Mulligan and D. F. McDonald, *Am. J. Phys.*, **25,** 180 (1957); J. H. Sanders, *The Fundamental Atomic Constants*, Oxford Univ. Press, New York, 1961; and J. H. Sanders, *The Velocity of Light*, Pergamon Press, Oxford, 1965.

[2]An even more precise direct comparison of the speeds for widely different wavelengths has been obtained from the study of flares (sudden outbursts) occurring in stars several light-years away. It has been found that radio waves with $\lambda = 1.2$ m arrive at the earth at the same time as visible light with $\lambda = 5.4 \times 10^{-7}$ m. The accuracy (a few parts in 10^7) of this comparison is limited chiefly by uncertainty about the mechanism of flare production. See B. Lovell, F. L. Whipple, and L. H. Solomon, *Nature*, **202,** 377 (1964).

crease of energy, even by many orders of magnitude, can do more than to narrow down the slight deficiency.

THE ENERGY-MOMENTUM RELATION FOR PHOTONS

Let us turn now to an important dynamical property of photons—the relation between energy and momentum. If a photon has energy E ($= h\nu$) it has an associated linear momentum of magnitude p such that

$$E = cp \tag{1–3}$$

The best (or certainly the most extensive) *experimental* support for this result is in a sense indirect. It is provided by the enormous body of data in nuclear and elementary-particle physics, where the analysis of collisions between individual photons and other particles is made by assuming Eq. (1–3) to hold.[1] Any inconsistency for that range of photon energies—from MeV to GeV[2]—would certainly have become apparent. The only experiments deliberately designed to test the energy-momentum relation have been made not with individual photons but with continuous beams of light, in studies of the *radiation-pressure phenomenon.* Such experiments involve the incidence of huge numbers of photons (for example, 1 watt of visible light represents a flow of about 3×10^{18} photons/sec) and can be adequately described and analyzed in terms of a steady flow of radiant energy, without reference to the photonic structure of the radiation. Indeed, the fact is that Eq. (1–3), as a general statement of the connection between energy and momentum for radiation in free space, was widely accepted long before the discovery of quantum behavior, because it was a necessary consequence of Maxwell's electromagnetic theory—the same theory that extracted the correct value of the speed of light from the physics of basic electric and magnetic phenomena. The radiation-pressure experiments were regarded primarily as a verification of Maxwell's theory. However, given a photon picture, they also imply that Eq. (1–3) holds for individual photons.

All radiation-pressure experiments are basically alike. They

[1]We shall consider such collisions in Chapter 6.

[2]1 GeV = 10^9 eV. This internationally adopted abbreviation (short for giga electron volt) is replacing BeV, which can cause ambiguity, because the European billion is 10^{12}, not 10^9.

consist in measuring the force F exerted on a surface by a known flux (measured by the incident power W) of radiant energy. The surface in question is a thin metal vane suspended on a delicate torsion fiber; the energy flux is measured by its heating effect. Account must be taken of the fraction ρ of the incident radiation that is reflected by the foil, because the reflected light contributes to the radiation force. (A perfectly reflecting surface would experience twice the force of a perfectly absorbing one for the same incident flux of radiation.)

The experiments almost always referred to in connection with radiation pressure are those of Nichols and Hull.[1] Certainly their measurements were very carefully and skillfully made, and their results verified Eq. (1–3) to better than 1%. But their work was done before the availability of high vacua in the laboratory, and they had to resort to special procedures to separate the true radiation pressure from a spurious *radiometer effect*. This is the phenomenon that an absorbing surface, becoming warmed by incident radiation, experiences a slight push from the gas adjoining it. The effect is very pronounced unless the vacuum is made extremely good—or, alternatively, quite bad! It can easily swamp the true radiation pressure and is the driving agency in the toy radiometers on sale in drugstores. Such radiometers always turn the wrong way, in fact, compared to what one would expect from true radiation forces. (Check this through your own observations if you have a chance to do so.) The very first quantitative experiment on radiation pressure (by a Russian, P. Lebedef, in 1901) was in fact done in a fairly good vacuum, but radiometer effects were nevertheless appreciable. The first really clean measurement of radiation pressure appears to have been achieved in a little-known investigation made in 1923 by Gerlach and Golsen.[2] Working with vacua better than 10^{-6} torr,[3] they rendered the radiometer effect inappreciable. The experiment can be regarded as a test of the following relation, arising from Eq. (1–3):

$$c = \frac{W(1 + \rho)}{F} \qquad (1\text{–}4)$$

using the quantities already defined. The right side of this equa-

[1] E. Nichols and G. F. Hull, *Phys. Rev.*, **13**, 307–320 (1901); **17**, 26–50, 91–104 (1903). Also G. F. Hull, *Phys. Rev.*, **20**, 292–299 (1905).

[2] W. Gerlach and A. Golsen, *Z. Physik.* (*Leipzig*), **15**, 1–7 (1923); A. Golsen, *Ann. Phys.*, **73**, 624–642 (1924).

[3] 1 torr = 1 mm of mercury at 0°C.

TABLE 1–3: RADIATION-PRESSURE EXPERIMENT

Material of vane	*Reflection coefficient* ρ	*Incident power* W, $\times 10^{-2}$ *watt*	*Measured force* F, $\times 10^{-10}$ *newton*	$W(1+\rho)/F$, $\times 10^{8}$ *m/sec*
Pt	0.60	6.07	3.14	3.09
Pt	0.60	2.80	1.44	3.11
Ni	0.43	6.39	3.23	2.83
Al	0.81	6.39	3.91	2.96
Al	0.81	2.78	1.74	2.89
			Av.	2.98

tion is made up of the quantities that were directly measured in the radiation-pressure experiment. W is, of course, the rate of arrival of energy and F is the rate of change of momentum of the radiation. It can then be tested whether this combination of W, ρ, and F is indeed equal to the speed of light. Table 1–3 shows the results of analyzing Golsen's data in this way. Thus the correctness of Eq. (1–3) is experimentally confirmed with an accuracy of about 2%.

The relation $E = cp$ for photons may be compared with the relation connecting kinetic energy, speed, and momentum for a particle in Newtonian mechanics. In the latter case we have $K = \frac{1}{2}vp$. Since the energy of a photon is all kinetic (for photons simply cease to exist when we try to stop them in an absorber), we might have been tempted to propose the relation $E = \frac{1}{2}cp$ (wrong!) for photons. This discrepancy might prompt one to ask what happens to the relation between kinetic energy and momentum for electrons as their speed is increased from relatively low values ($< 0.1c$, say) up to values about equal to c. The answer is that at low energies ($K < 1$ keV) the relationship is fairly accurately Newtonian,[1] but that at high energies ($K > 0.1$ MeV) the momentum becomes significantly less than one would calculate from the value of $2K/v$, and at very high energies ($K > 50$ MeV) is given by K/c with an accuracy of better than 1%. (The evidence for this is to be found in a study of atomic collisions involving energetic electrons; we shall say more about such processes later.) Thus, just as with the relation between kinetic energy and speed, one sees a smooth but unmistakable departure from Newtonian behavior when sufficiently high energies, and speeds approaching that of light, are involved.

[1]See the film, *Momentum of Electrons*, by J. G. King, Education Development Center, Newton, Mass., 1963.

This serves to reinforce our belief that the dynamics of photons and of other particles can be brought, for some purposes at least, within the same descriptive framework. Our next step will be to suggest what that framework might be. Our argument will appeal to one's sense of what is plausible; it will not be logically inescapable. But as the old saying goes, "the proof of the pudding is in the eating," and we shall see how beautifully one can describe the transition from Newtonian to non-Newtonian behavior on the basis of our conclusions (which are indeed precisely those of special relativity).

MATTER AND RADIATION: THE INERTIA OF ENERGY

Are not gross Bodies and Light convertible into one another, and may not Bodies receive much of their Activity from the Particles of Light which enter their Composition?

Newton, *Opticks* (4th ed., 1730)

It would be quite wrong to suggest that Newton had really anticipated 20th-century physics to the extent that the above quotation might imply, but his provocative query is superbly appropriate as an introduction to the discussion that we shall now undertake. For we shall consider the intimate connection between the inertia of ordinary matter and the energy of radiation, and in so doing we shall develop some dynamical results that apply equally to photons and "gross bodies." We shall obtain, as one of the consequences, a full account of the relation between speed and kinetic energy for the electrons in the ultimate-speed experiment.

Our starting point will be a *gedanken experiment* (literally a "thought experiment," i.e., a fictitious, not really feasible experiment) which was invented by Einstein himself in 1906.[1] The purpose of it is to suggest that energy must have associated with it a certain inertial mass equivalent.[2] We suppose that an amount E of radiant energy (a burst of photons) is emitted from one end of a box of mass M and length L that is isolated from its surroundings and is initially stationary [Fig. 1–4(a)]. The radiation carries momentum E/c. Since the total momentum of the system remains equal to zero, the box must acquire a momentum equal

[1] A. Einstein, *Ann. Phys.*, **20**, 627–633 (1906).

[2] By inertial mass we mean the ratio of linear momentum to velocity.

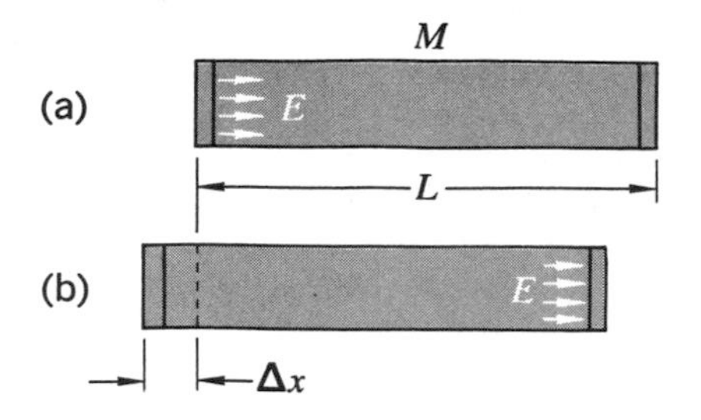

Fig. 1–4 Einstein's box—a hypothetical experiment in which a box recoils from its initial position (a) to a final position (b) as a result of a burst of radiant energy traveling from one end of the box to the other.

to $-E/c$. Hence the box recoils with a speed v, given by

$$v = -\frac{E}{Mc} \tag{1–5}$$

After traveling freely for a time Δt ($= L/c$ very nearly, provided $v \ll c$), the radiation hits the other end of the box and conveys an impulse, equal and opposite to the one it gave initially, which brings the box to rest again.[1] Thus the result of this process is to move the box through a distance Δx:

$$\Delta x = v\,\Delta t = -\frac{EL}{Mc^2} \tag{1–6}$$

But this being an isolated system, we are reluctant to believe that the center of mass of the box plus its contents has moved. We therefore postulate that the radiation has carried with it the equivalent of a mass m, such that

$$mL + M\,\Delta x = 0 \tag{1–7}$$

Putting the last two equations together, we have

$$m = \frac{E}{c^2} \quad \text{or} \quad E = mc^2 \tag{1–8}$$

For the man on the street, Einstein and relativity are probably epitomized by this result. For the physicist, its importance is not lessened by its becoming hackneyed; it asserts a fundamental inertia of energy. Although the calculation as we have presented it (which differs somewhat from Einstein's original version) points in the first instance to the mass associated with radiant energy, one quickly recognizes that the implications are much wider than this. When the radiation is emitted from one end of Einstein's box, that end must surely suffer a decrease, by

[1]If you feel that more careful account should be taken of the recoil of the box and its effect on the time and distance of transit of the radiation, see Problem 1–13.

the amount E/c^2, in its inertial mass. Likewise, the absorption of the radiation at the other end means an addition to the mass of that portion. Once the energy has been absorbed, it loses its identification as the energy of photons and ultimately becomes just an addition to the thermal energy. And we are quickly led to the idea that energy in any form has the mass equivalent defined by Eq. (1–8)—a general principle of the inertia of energy.[1]

The prime example of the mass-energy equivalence, to which we owe our continuing existence, is provided by thermonuclear reactions occurring in stars such as the sun. Observation tells us that radiant energy is reaching us from the sun at the rate of 1.35×10^3 watts/m^2. Given this figure and Eq. (1–8), we can infer that the mass of the sun is decreasing at the rate of about 4.5×10^6 tons/sec—an impressively rapid loss, even though it is only about 1 part in 10^{13} of the sun's mass per year. This comes about through chains of nuclear reactions, chief among which is the sequence by which hydrogen (^{1}H) is converted to helium (^{4}He). One must, of course, have four hydrogen atoms to end up with one helium atom, and the process takes place in several separate steps. One of these steps is particularly worth mentioning here, because it is a simple and remarkably direct example of the equivalence of the mass of ordinary matter and the energy of photons. It is this:

$$p + \mathrm{D} \rightarrow {}^3\mathrm{He} + \gamma \qquad (1\text{–}9)$$

A proton fuses with a deuteron D (the nucleus of hydrogen-2, containing one proton and one neutron), making a system of two protons and one neutron, which is the nuclear composition of ^{3}He. But, as mass-spectrometer measurements show us, the mass of this combination is greater than the mass of ^{3}He in its normal state. Here are the approximate values:

Proton	1.6724×10^{-27} kg
Deuteron	3.3432
$p + \mathrm{D}$	5.0156
^{3}He nucleus	5.0058
Mass excess	9.8×10^{-30} kg

This amount of mass is carried off by a photon (a γ ray) as indicated by Eq. (1–9). The energy of that photon is given by

[1]For a fine discussion of this question, see M. von Laue's article "Inertia and Energy" in *Albert Einstein: Philosopher-Scientist*, Vol. II, (P. A. Schilpp, ed.), Harper Torchbook, Harper and Row, New York, 1959.

Eq. (1–8):

$$E = mc^2 = 9.8 \times 10^{-30} \times 9.0 \times 10^{16}$$
$$= 8.8 \times 10^{-13} \text{ joule}$$
$$= 5.5 \text{ MeV}$$

This process has been studied in the laboratory, and γ rays of the expected energy have been observed.[1] It should perhaps be added that such reactions, when they occur as thermonuclear reactions in the sun, require temperatures of the order of 10^7 °K and thus take place only in the inner regions. Gamma rays, such as those just considered, are completely absorbed before reaching the sun's surface, and their energy finally escapes in photons with individual energies of the order of only 1 eV—infrared, visible, and ultraviolet—that constitute the familiar solar spectrum.

The equation $E = mc^2$ has (at least in popular accounts) been so exclusively linked to nuclear transformations as to divert attention from its universality. But the message of Einstein's equation is that *any* change ΔE in the energy of a body implies a corresponding change Δm in its inertial mass:

$$\Delta E = c^2 \Delta m \qquad (1\text{–}10)$$

A golf ball in motion has more mass than the same golf ball at rest. The heated filament of a lamp has more mass than the same filament when cold. A charged capacitor has more mass than the same capacitor uncharged. And so on. Because, in terms of familiar magnitudes, the mass associated with a given amount of energy is exceedingly small (e.g., the energy used per day for domestic purposes in a city of a million people has a mass equivalent of only about 1 g), this intimate connection between the two was long unrecognized. Einstein regarded the discovery of this connection as being extremely important. To quote his own words[2]:

> The most important result of a general character to which the special theory has led is concerned with the conception of mass. Before the advent of relativity, physics recognized two conservation laws of fundamental importance, namely, the law of the conservation of energy and the law of the conservation of mass; these two fundamental laws appeared to be quite inde-

[1] W. A. Fowler, C. C. Lauritsen, and A. V. Tollestrup, *Phys. Rev.*, **76,** 1767 (1949).

[2] A. Einstein, *Relativity*, Crown, New York, 1961.

pendent of each other. By means of the theory of relativity they have been united into one law.

Perhaps one of the best ways to appreciate the pervasive character of the mass-energy equivalence is to consider a single, neutral atom in a piece of ordinary matter. From one point of view it is just one of a collection of what Newton called "solid, massy, hard, impenetrable, movable Particles."[1] The question of any inner structure does not arise, and it seems almost obvious that the atom's inertial property should be described by a single quantity that we call the mass. But now consider this same atom from the standpoint of present-day knowledge. It is a complicated assembly of electrons, neutrons, and protons (and if we want to probe more deeply, there is finer structure yet). The mass of the atom as a whole contains positive contributions from the kinetic energies of its swiftly moving constituents, and contributions of both signs (predominantly negative) from the potential energy of their electrical and nuclear interactions. (Note that a force of attraction between two particles automatically represents a *negative* contribution to the total mass of the system.[2]) Any change in the internal state of the atom is accompanied by a flow of energy into or out of it, with an associated increase or decrease in its mass. The ability of the constituents to cohere depends on the fact that their total energy in this configuration is less than if they were all separated from one another. In these terms, then, the mass of an atom is the result of a remarkable and subtle synthesis. Yet it serves to characterize the whole atom in every dynamical context—including gravitation—in which it moves as a single unit.

ENERGY, MOMENTUM, AND MASS

Let us now try to put together some of the results we have discussed. For photons we have

$$E = cp \tag{1–3}$$

and

$$m = \frac{E}{c^2} \tag{1–8}$$

[1]Sir I. Newton, *Opticks*, 4th ed., 1730; reprinted in revised form by G. Bell, London, 1931; Bell edition reprinted by Dover, New York, 1952.

[2]Provided the strength of the attractive force gets less with increasing separation, which is true of all such forces between elementary particles in atoms.

(the first experimental, the second based on Einstein's box). Combining these, we have

$$m = \frac{p}{c} \tag{1-11}$$

In Newtonian mechanics, however, we have

$$m = \frac{p}{v} \tag{1-12}$$

It looks as though we might regard Eq. (1–11) as a particular case of Eq. (1–12), for $v = c$. If, further, we suppose that Eq. (1–8) describes a universal equivalence of energy and inertial mass, we can combine Eqs. (1–8) and (1–12) into a single statement:

$$E = \frac{c^2 p}{v} \tag{1-13}$$

Now in classical mechanics we are never concerned with absolute energies but only with energy differences, and with the transformation between one form of energy and another. A particle suffers a change of potential energy, for example, and its kinetic energy undergoes a corresponding change, so that the total energy remains constant. The basis for analyzing all such situations is *Newton's law.* The increment of kinetic energy corresponds to the work done by external forces,[1] and we have

$$dE = F\,dx = \frac{dp}{dt}\,dx$$

i.e.,

$$dE = v\,dp \tag{1-14}$$

If we accept Eqs. (1–13) and (1–14) we can obtain from them a relationship, now proposed as a general one, between energy and momentum for a particle. We do this by multiplying together the left and right sides of the two equations, and integrating:

$$E\,dE = c^2 p\,dp$$

Therefore,

$$E^2 = c^2p^2 + E_0{}^2 \tag{1-15}$$

where $E_0{}^2$ is a constant of integration, written explicitly as the square of some constant energy.

[1]The ultimate-speed film presents evidence that, even under conditions where some of the features of Newtonian mechanics have broken down, the increase of energy (kinetic energy) of an electron is still equal to the work calculated from the electrostatic force multiplied by the distance traveled.

From here it is possible to proceed in several ways. For example, we can substitute in Eq. (1–15) the relation $cp = Ev/c$ from Eq. (1–13). This leads at once to the following result:

$$E(v) = \frac{E_0}{(1 - v^2/c^2)^{1/2}} \tag{1–16}$$

For $v \ll c$ we can approximate this exact result by the binomial expansion, neglecting terms of higher order than v^2/c^2.

$$\text{[Approximate result } (v \ll c)] \quad E(v) \approx E_0 + \tfrac{1}{2}\left(\frac{E_0}{c^2}\right)v^2 \tag{1–17}$$

If Eq. (1–17) is to harmonize with Newtonian mechanics at low velocities, we must identify E_0/c^2 with the classical inertial mass of a particle: Let us denote this by m_0. Then Eqs. (1–8) and (1–16) together lead to an explicit variation of inertial mass with speed:

$$m(v) = \frac{m_0}{(1 - v^2/c^2)^{1/2}} \tag{1–18}$$

The quantity m_0, which in Newtonian mechanics would be *the* inertial mass of a body, now assumes a new role as the *rest mass* of the body for $v = 0$; at any other speed the inertial mass is greater.[1]

An increase of inertial mass with speed is of course implied as soon as one embraces a general principle of the inertia of energy. The particular form of variation expressed by Eq. (1–18) is shown graphically in Fig. 1–5, together with some experimental results based on the electric and magnetic deflection of energetic electrons.

Equations (1–15) and (1–18) are two of the central results of the new dynamics; the first of them—the relation between energy and momentum—will prove to be of special importance and applicability. But the kinetic energy of a particle, so valuable a quantity in classical dynamics, now takes on a secondary status. It is merely the difference between the total energy E and the rest energy E_0:

$$K = m_0c^2\left[\frac{1}{(1 - v^2/c^2)^{1/2}} - 1\right] \tag{1–19}$$

Of course K remains a quantity of practical importance, because it is the measure of the extra energy conferred on a particle

[1]And the quantity E_0 ($= m_0c^2$) is the *rest energy*. Thus for electrons (for example) we have $m_0 = 9.11 \times 10^{-31}$ kg, $E_0 = 8.2 \times 10^{-14}$ joule = 0.51 MeV.

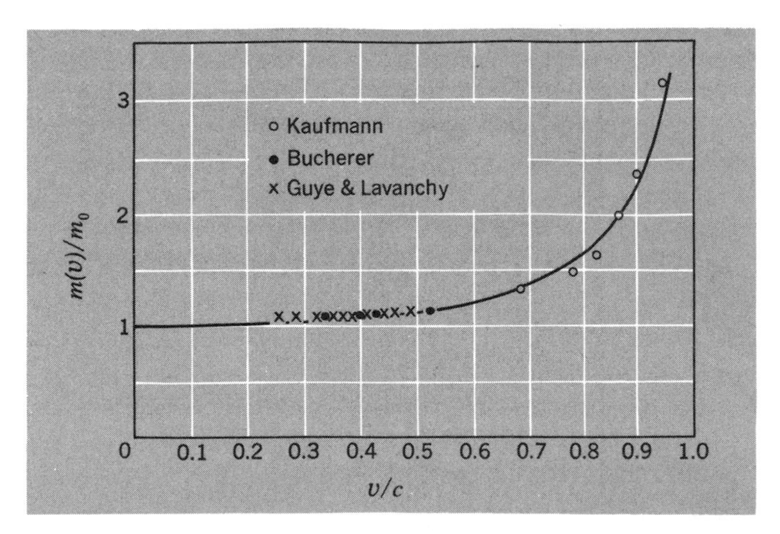

Fig. 1–5 Variation of inertial mass with speed for electrons. Based on data of Kaufmann (1910), Bucherer (1909), and Guye and Lavanchy (1915). (After R. S. Shankland, Atomic and Nuclear Physics, Macmillan, New York, 1961.)

through the work done by external forces. Note that K is *not* obtained by substituting into the expression $\frac{1}{2}mv^2$ the value of m calculated from Eq. (1–18)—a frequently made error, because the temptation to cling to the Newtonian form of the kinetic energy is very strong.

At the risk of seeming repetitious, let us reemphasize the significance of $m(v)$ as defined by Eq. (1–18). It describes the inertial property of a body moving with velocity $\mathbf{v}$, so that the momentum $\mathbf{p}$ is given by the equation

$$\mathbf{p} = m(v)\mathbf{v} \tag{1–20}$$

It also describes the total energy content of the body, so that

$$E = m(v)c^2 \tag{1–21}$$

Now it is the quantities $\mathbf{p}$ and E, rather than $m(v)$ by itself, that figure in any actual dynamical situation. In this sense the variable mass $m(v)$ is just a convenient construct which, for example, allows us to preserve the form of the Newtonian statement that momentum is mass times velocity. Many physicists prefer to reserve the word *mass* to describe the rest mass m_0, a uniquely defined property of a given particle, But this is essentially a matter of taste.[1] Whatever words one elects to use, there is no disagreement on the fact that Eqs. (1–20) and (1–21) describe the momentum and total energy of a particle, where $m(v)$ is given by Eq. (1–18).

[1]And one cannot escape the fact that, for almost any particle, even the rest mass involves contributions associated with the motions and kinetic energies of its constituents.

The denominator $(1 - v^2/c^2)^{1/2}$ appears so often in special relativity, and is so awkward to write, that nearly all discussions of relativity make use of a single symbol, γ, defined as follows. Put

$$\gamma(v) = \frac{1}{(1 - v^2/c^2)^{1/2}} \tag{1–22}$$

Then we have

$$m = \gamma m_0 \tag{1–23}$$

$$\mathbf{p} = \gamma m_0 \mathbf{v} \tag{1–24}$$

$$E = \gamma m_0 c^2 \tag{1–25}$$

where in using Eqs. (1–23) to (1–25) we must remember that γ depends on the speed v according to Eq. (1–22).

IS THE NEW DYNAMICS CORRECT?

It is important to ask whether Eq. (1–19) does indeed provide a correct account of the relation between speed and kinetic energy as observed, for example, in the linac experiment. Rearranging the result, we have

$$1 + K/m_0c^2 = (1 - v^2/c^2)^{-1/2}$$

Therefore,

$$1 - v^2/c^2 = (1 + K/m_0c^2)^{-2}$$

or

$$v^2 = c^2[1 - (1 + K/m_0c^2)^{-2}] \tag{1–26}$$

Clearly the rest energy m_0c^2 provides a natural unit in which to measure the extra energy K that is added to a particle by means of an acceleration process. We can, in fact, draw up a table showing how the speed would depend on K for any particle whatsoever (Table 1–4).

Given that, for electrons, $m_0c^2 = 0.51$ MeV, we can readily plot a curve of v^2 in m^2/sec^2 against K in MeV. This curve has been drawn in on the graph of the data in the ultimate-speed experiment (Fig. 1–3). It may be seen that the agreement between theory and experiment is very good, and speaks strongly for the correctness of the revised dynamics, as does the measured variation of mass with speed, shown in Fig. 1–5.

If we wanted to plot a curve of v^2 versus K for protons, all

TABLE 1-4: SPEED VERSUS KINETIC ENERGY FOR PARTICLES

K/m_0c^2	$(1 + K/m_0c^2)^{-2}$	v^2/c^2	v/c	v^2, $\times 10^{16}$ m^2/sec^2
0.1	0.8264	0.1736	0.417	1.56
0.2	0.6944	0.3056	0.553	2.75
0.3	0.5917	0.4083	0.639	3.67
0.5	0.4444	0.5556	0.745	5.00
1.0	0.2500	0.7500	0.866	6.75
2.0	0.1111	0.8889	0.943	8.00
5.0	0.0278	0.9722	0.986	8.87
10.0	0.0083	0.9917	0.996	8.93
30.0	0.0010	0.9990	0.999	8.99

we would need to do would be to put $m_0 = 1.672 \times 10^{-27}$ kg, which gives $m_0c^2 = 0.938$ GeV (or 938 MeV), and Table 1-4 would provide the rest of the information needed. The fact that this does indeed give correct results for protons is amply attested in the operation of big nuclear accelerators, and there is plenty of evidence that Eq. (1-26) holds for particles of all kinds.

Among the various features of these modified laws of motion, the phenomenon of the limiting speed c is perhaps the most noteworthy. It means that energy (and mass) can be piled onto atomic particles without increasing their speed appreciably. To see in detail how this works, it is convenient to rewrite Eq. (1-16) as follows:

$$(1 - v^2/c^2)^{1/2} = E_0/E$$

Therefore,

$$v^2/c^2 = 1 - (E_0/E)^2 \qquad (1\text{-}27a)$$

and

$$v/c = [1 - (E_0/E)^2]^{1/2}$$

For $E \gg E_0$, we then have, approximately,

$$v/c \approx 1 - \tfrac{1}{2}(E_0/E)^2 \qquad (1\text{-}27b)$$

For example, the Harvard-M.I.T. electron accelerator has as its injector a linear accelerator (like the one used in the ultimate-speed film) that gives the electrons 15 MeV energy. The main accelerator brings the electrons up to about 5 GeV (= 5000 MeV).

Using these values, one finds

Injection from linac (15 MeV) $\rightarrow v/c \approx 0.9995$

Final energy (5 GeV) $\rightarrow v/c \approx 0.99999995$

Thus the change of v/c after the preliminary acceleration is only about 5 parts in 10^4. These big nuclear machines might appropriately be called "ponderators"[1] rather than accelerators, for to an excellent approximation they do just add mass to the particles injected into them, with no significant increase in the speed as such.

MOTION UNDER A CONSTANT FORCE

The simplest dynamical problem in classical mechanics is the motion of a body under a constant force. Let us see how this problem is modified in the new dynamics. Suppose a force F acts on a body for a time t (we assume one-dimensional motion); the body is assumed to be initially at rest, and ends up with a speed v. Then

$$Ft = mv = \frac{m_0 v}{(1 - v^2/c^2)^{1/2}} \tag{1-28}$$

Therefore,

$$1 - v^2/c^2 = (m_0 v/Ft)^2$$
$$c^2 = v^2[1 + (m_0 c/Ft)^2]$$

and

$$v(t) = \frac{c}{[1 + (m_0 c/Ft)^2]^{1/2}} \tag{1-29}$$

This is a rather complex-looking result. Let us consider two extreme cases:

(a) $Ft \ll m_0 c$:

$$(m_0 c/Ft)^2 \gg 1$$

Therefore,

$$v(t) \approx \frac{c}{(m_0 c/Ft)} = \frac{F}{m_0} t$$

(b) $Ft \gg m_0 c$:

$$(m_0 c/Ft)^2 \rightarrow 0$$

[1]This name was first proposed around 1945 by Prof. A. G. Hill of M. I. T.

Therefore,

$$v(t) \approx c$$

Case (a) corresponds to ordinary Newtonian mechanics. Case (b) displays the now-familiar property of a limiting constant speed c for motion under any force, no matter how large it is or for how long it is applied.

"EINSTEIN'S BOX UNHINGED"

According to our present beliefs as expressed by special relativity, the speed of light in free space represents an upper limit, not only to the speed of material particles such as electrons, but also to the speed with which an interaction of any kind can be propagated—gravitational, nuclear, electric, etc. Were this not so, it would be possible (as we shall discuss later) to arrive at a paradox involving the interchange of the roles of cause and effect, according to one's point of view (see the discussion of causality near the end of Chapter 4).

One particular consequence of the physical speed limit equal to c is that the classical concept of an ideal rigid body finds no place in special relativity. (And strictly speaking, it cannot be justified in classical mechanics either.) For by a rigid body we mean an object along which physical information can be transmitted in an arbitrarily short time, so that the object is set in motion instantaneously, as a single unit, when a force is applied to any point in it. For any ordinary box the information that one end has been struck is transmitted as an elastic wave, which we know is many orders of magnitude slower than a light signal. Thus the Einstein box argument in its original form cannot be maintained. At the receiving end of the box, the first intimation that anything had happened at the other end would be the arrival of the radiation itself. We can, however, rehabilitate the argument as follows.

Ignore completely any connection between the ends of the box, and regard it as two separate masses, m_1 and m_2 (Fig. 1–6). Just suppose that one end, of initial mass m_1, emits energy E at $t = 0$ and suffers a mass change to m_1'. It acquires a velocity v_1 given by

$$v_1 = \frac{-E/c}{m_1'}$$

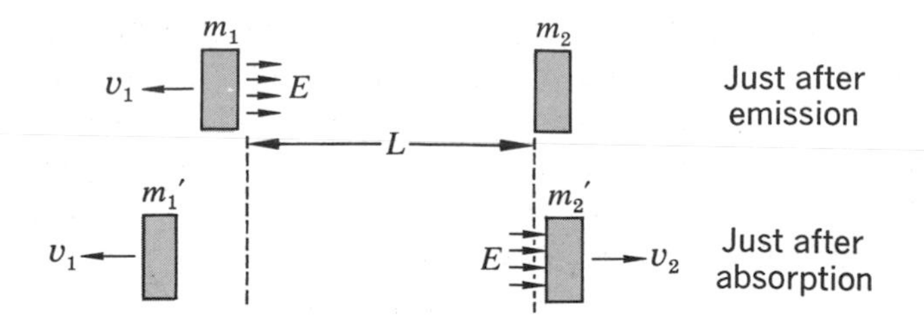

Fig. 1–6 "Einstein's box unhinged." The recoil processes in two unconnected masses in consequence of a burst of radiant energy emitted from one (m_1) and absorbed in the other (m_2).

If m_1 were originally at $x = 0$, its position at any later time is thus given by

$$x_1(t) = -\frac{E}{m_1'c}t \tag{1–30}$$

When the energy arrives at m_2 (at $t = L/c$) it causes a recoil and a change of mass so that we have, for the position of m_2,

$$x_2(t) = L + \frac{E}{m_2'c}(t - L/c) \tag{1–31}$$

Let the total mass be M, and let the position of the center of mass be $\bar{x}$ before the radiation was emitted from m_1 and $\bar{x}'$ after it was absorbed in m_2. Then

$$M\bar{x} = m_1 \cdot 0 + m_2 \cdot L \tag{1–32}$$

and

$$M\bar{x}' = m_1'\left(\frac{-E}{m_1'c}t\right) + m_2'\left[L + \frac{E}{m_2'c}(t - L/c)\right]$$

i.e.,

$$M\bar{x}' = -\frac{E}{c}t + m_2'L + \frac{E}{c}t - \frac{E}{c^2}L \tag{1–33}$$

Hence, if $\bar{x}' = \bar{x}$,

$$\Delta m_2' = m_2' - m_2 = \frac{E}{c^2} = -\Delta m_1' \tag{1–34}$$

Thus the principle of inertia of energy finds a sounder theoretical basis, but by this stage we have seen its real vindication in the experimentally observed behavior of particles.

SOME COMMENTS

In this chapter we have presented evidence to show that the behavior of particles at very high speed simply does not conform to Newtonian dynamics. By analyzing this behavior, and by following Einstein in the assumption that the center of mass of an isolated system does not spontaneously shift, we have developed some relations (which appear experimentally to be valid for all attainable speeds) connecting energy, momentum, and mass. This has allowed us to arrive rather quickly at some important dynamical results. On the other hand, it is clear that the arguments we have used involve a good deal of conjecture; they are suggestive but by no means irresistible. Furthermore, one may well ask what all this has to do with the things one normally thinks of when relativity is mentioned—such things as the Lorentz contraction, frames of reference, space-time, the Michelson-Morley experiment. The answer is that the connection is very, very close. But apart from one small hint in our discussion of the results of the ultimate-speed experiment, we have so far not tried to deal with these very fundamental aspects of relativity. There is a good reason for that; each of the experiments that we cited was conducted within a single frame of reference—the experimenter's laboratory. But the concepts of distance, time, and velocity were involved at every turn; without them it is impossible to formulate or discuss dynamics.

It was in the attempt to explain optical phenomena that the need for some drastic revision of our ideas about space and time finally became overwhelming. The development of this problem, culminating in the Michelson-Morley experiment, is the subject of Chapter 2. And then we shall see how Einstein, through his insistence on a fundamental reexamination of the bases of dynamical measurement, made it possible to fit everything together within a single dynamical scheme. The same concepts of space and time are found to be appropriate to the facts of optics and electromagnetism and to the non-Newtonian dynamical behavior that we have been discussing in this chapter. Our program, then, will be to describe the predicament engendered by the facts of optics, to show how Einstein eliminated the apparent conflict between optics and Newtonian mechanics, and then to illustrate some of the applications of Einstein's formulation of the principle of relativity.

PROBLEMS

1-1 A burst of 10^{14} electrons accelerated to an energy of 15 MeV per electron is stopped in a copper target block of mass 100 g. If the block is thermally insulated, what is its temperature rise? The specific heat of copper is 0.09 cal/g·°K.

1–2 The intensity of the sun's radiation just outside the earth's atmosphere is approximately 8×10^4 joules/m^2·min.

(a) Approximately what force does this radiation exert on the Echo II reflecting satellite balloon? Echo II is a spherical shell of radius 20.4 m. Its skin consists of a layer of Mylar plastic, 9×10^{-6} m thick, between two layers of aluminum, each 4.5×10^{-6} m thick. The density of Mylar is 10^3 kg/m^3; of aluminum 2.7×10^3 kg/m^3.

(b) Compare this force with the sun's gravitational force on the balloon.

1–3 (a) Radiant energy from the sun is received at the earth at the rate of about 2 cal/cm^2 · min on a surface perpendicular to the sun's rays. What total force would be exerted on the whole earth by solar radiation if it were all absorbed? How does this compare with the sun's gravitational force on the earth?

(b) What radius would a particle of dust in space have to have to be in equilibrium under the combined effects of the sun's gravitational attraction and radiational repulsion?

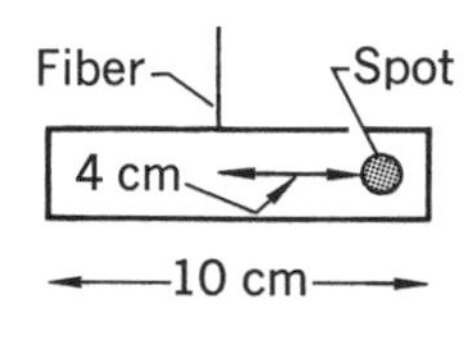

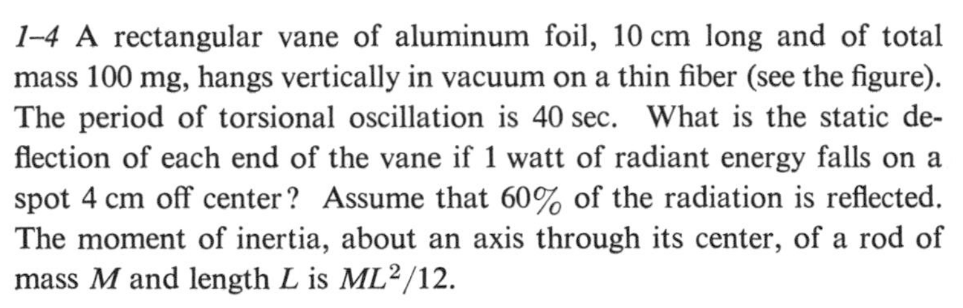

1–4 A rectangular vane of aluminum foil, 10 cm long and of total mass 100 mg, hangs vertically in vacuum on a thin fiber (see the figure). The period of torsional oscillation is 40 sec. What is the static deflection of each end of the vane if 1 watt of radiant energy falls on a spot 4 cm off center? Assume that 60% of the radiation is reflected. The moment of inertia, about an axis through its center, of a rod of mass M and length L is $ML^2/12$.

1–5 It has been said that a fully opened umbrella catches about enough radiant energy per second on a clear day to run a washing machine. Use this statement as a basis for calculating the approximate rate of loss of mass by the sun.

1–6 If all the light used in New York City in 1 hour of the evening could be captured and put in a box, approximately how much heavier would the box become?

1–7 A battery connected to a flashlight bulb is exactly counterpoised on the pan of a balance. The battery maintains an average current of 0.1 amp at an average voltage of 1 volt for 3 hours. Assuming all this energy is radiated away, what must be the order of magnitude of the sensitivity $\Delta M/M$ of the balance if a deflection is to be detected?

1–8 (a) Sir Arthur Eddington once remarked that if 1 g of electrons could be confined in a sphere of 10 cm radius, the mass associated with their electric potential energy would be of the order of 10 million tons. Check this assertion for yourself, assuming that the electrons form a ball of charge of uniform density. (The electrostatic potential energy of a sphere of charge q and radius r is $3kq^2/5r$, where q is the charge in coulombs, r the radius in meters, and k—the constant in the inverse-square law $F = kq_1q_2/r^2$—is equal to 9×10^9 newton·m^2/coulomb^2.)

(b) The calculation in (a) may sound pretty outlandish. But now calculate the mass of all the electrons in a sphere of water of 10 cm radius. If it were not for the positively charged nuclei, we should be faced with almost exactly such a situation as Eddington describes.

1–9 An eccentric billionaire decides to sterilize his 10^6-liter swimming pool by boiling the water in it. For heating purposes he uses the fusion reaction

$$^1\text{H} + {}^3\text{H} \rightarrow {}^4\text{He} + \text{radiant energy}$$

Assuming the heating system is 20% efficient, how much does he pay for the tritium (^{3}H) to raise the pool temperature from 20 to 100°C? It takes 4.2 joules to raise 1 g of water through 1°C. Tritium costs about $5 per cm^3 of gas at STP.

Atomic masses:	^{1}H	1.0081 amu
	^{3}H	3.0170 amu
	^{4}He	4.0039 amu

1–10 A spherical nuclear reactor of mass 10^{30} kg in interstellar space is completely surrounded by a thin, nonrigid spherical shell of matter with a mass of 10^{26} kg (see the figure). The reactor loses 10^{10} kg/sec of its mass by the emission of electromagnetic radiation. (If this sounds like the description of a star, it's no accident!) This radiation is completely absorbed by the surrounding shell of matter. What must be the radius of the spherical shell if the repulsion exerted on it by the radiation is just great enough to balance the gravitational attraction exerted on it by the reactor? (Consider the forces exerted on a small portion of the shell.)

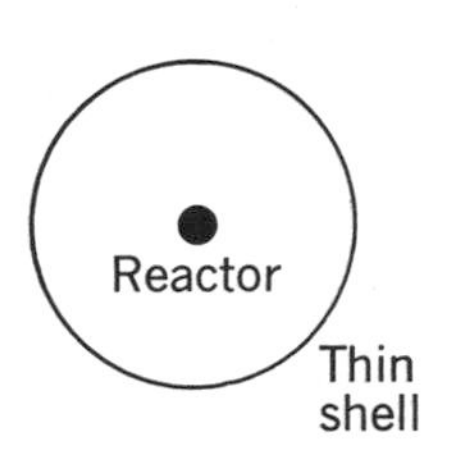

1–11 (a) The heat of formation of CO from C and O is about 20 kcal/mole. If this heat is allowed to escape, by what fraction is the carbon monoxide lighter than its parent elements?

(b) If the nuclei of the abundant isotopes of C and O (^{12}C and ^{16}O) could be combined to produce ^{28}Si in its normal state, with the escape of all surplus energy, by what fraction would the silicon nucleus be lighter than its parent nuclei?

(c) Process (b) would not be very probable unless the reacting nuclei could be forced to within about 10^{-14} m against their electric repulsion. What kind of temperature would be needed in a mixture of C and O before this became at all likely? In the Coulomb force law ($F = kq_1q_2/r^2$) the value of k is 9×10^9 newton·m^2/coulomb^2.

1–12 (a) A body of mass $m_1 + \Delta m$ is connected to a body of mass $m_2 - \Delta m$ by a spring of spring constant k and negligible mass (see the figure). The system is at rest on a frictionless table. A burst of radiation is emitted by the first body and absorbed by the second, changing the masses to m_1 and m_2 and setting the system into oscillations. If the time of transit of the radiation is negligibly small compared to the period of oscillation, show that the maximum extension of the spring is given by

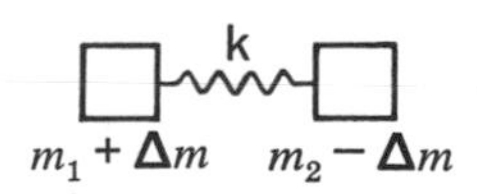

$$x = c\,\Delta m \left(\frac{m_1 + m_2}{m_1 m_2 k}\right)^{1/2}$$

(b) Consider qualitatively what would happen if the spring could be made stronger and stronger, without limit.

1–13 The discussion of Einstein's box in the text assumes that the transit time of the radiant energy is L/c (where L is the length of the box), a result obtained by neglecting the distance of recoil of the box. The text also ignores the decrease in the mass of the box resulting from the emission of the radiant energy. Show that if both these features are properly taken into account, the result $m = E/c^2$ is still obtained.

1–14 Having obtained the relation $m = E/c^2$ for the inertial mass of radiant energy, Einstein in 1911 speculated whether this same value of m, substituted in the universal gravitation formula, would describe the deflection by the sun of light rays from a distant star, thereby causing the apparent direction of the star to be slightly displaced. (A German astronomer, J. Soldner, treating light simply as Newtonian particles traveling at speed c, had—unknown to Einstein—carried out essentially this same calculation back in 1801!)

Calculate the deflection α for a photon that just grazes the edge of the sun, by assuming that to a first approximation it shoots by along a straight-line path always traveling at speed c, but that the component of the gravitational force perpendicular to the path ($F\cos\theta$), in-

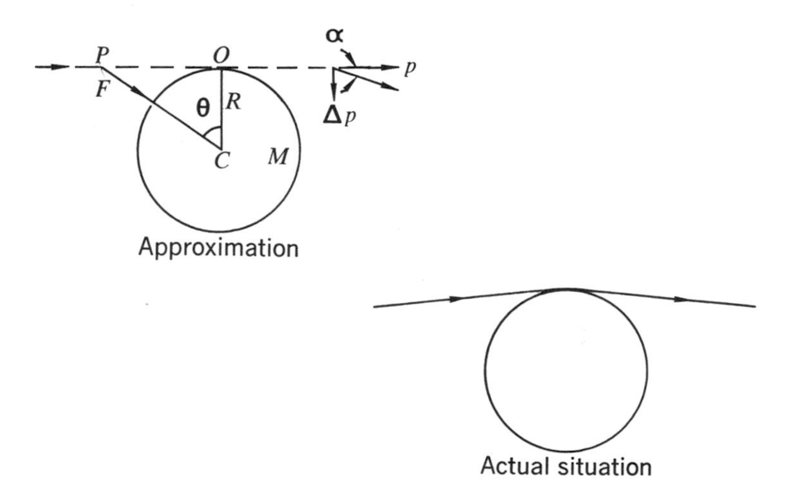

tegrated over the complete path, ends up by giving the photon a transverse momentum component

$$\Delta p = \int F \cos\theta \, dt$$

so that the deflection α is equal to $\Delta p/p$, where $p = E/c$ (see the figure). Let x be the distance measured along the path. Taking the origin of x at the point where the ray grazes the sun, the limits of x are $\pm\infty$ and we also have $dt = dx/c$, and $CP^2 = x^2 + R^2$.

The result is $\alpha = 2GM/c^2R$. After verifying this formula, put in numbers to obtain a numerical value of α in seconds of arc. (*N.B.:* It is believed that this answer is wrong in principle. Einstein's general theory of relativity brings in a further factor of 2 in the theoretical deflection.)

1–15 A particle is given a kinetic energy equal to n times its rest energy m_0c^2. What are

(a) Its speed?

(b) Its momentum?

1–16 (a) Through what voltage would an electron have to be accelerated from rest so as to increase its mass by 0.4%?

(b) What would be its speed under these conditions?

1–17 A proton is accelerated through the equivalent of 500 million volts by a synchrotron.

(a) What is its mass, expressed as a multiple of the rest mass?

(b) What fraction is its speed of the speed of light?

1–18 What is the velocity of the center of mass of a system consisting of a photon of energy $h\nu$ and a stationary atom of rest mass m_0? How would your answer change if the atom were in an excited state? Does this seem reasonable nonrelativistically?

1–19 (a) The ratio v/c is very often denoted by the single symbol β. Show that if $\beta \ll 1$, the following are valid through terms of order β^2:

$$E = m_0c^2 + m_0v^2/2 = m_0c^2(1 + \beta^2/2)$$

$$K = m_0v^2/2 = m_0c^2\beta^2/2$$

$$pc = m_0vc = m_0c^2\beta$$

$$\gamma = 1 + \beta^2/2$$

(b) Show that if $\gamma = \epsilon^{-1} \gg 1$, the following are valid through terms of order ϵ^2:

$$\beta = 1 - \epsilon^2/2$$

$$E = \epsilon^{-1}m_0c^2$$

$$K/E = 1 - \epsilon$$

$$pc/E = 1 - \epsilon^2/2$$

$$K/pc = 1 - \epsilon + \epsilon^2/2$$

1–20 (a) The refractive index of air for visible light is about 1.000277. What would be the kinetic energy of an electron whose speed through an evacuated tube would be sufficient to keep it neck-and-neck with a light signal traveling through the air?

(b) What force would a stream of such electrons, equivalent to a current of 16 mA, exert on a block of material in which they were stopped?

1–21 (a) What fractional error does one make in using $\frac{1}{2}Mv^2$ for the kinetic energy of a body if its speed is

(1) 3 m/sec
(2) 300 m/sec (the speed of sound in air)
(3) 10^4 m/sec (the speed a body needs to escape from the earth)
(4) $0.1c$
(5) $0.9c$

(b) If the experimental error in a measurement of v is 1% (approximately), how large must v be before relativistic corrections become significant?

1–22 An electron moving with a speed $0.5c$ in the x direction enters a region of space in which there is a uniform electric field in the y direction. Show that the x component of the velocity of the particle must decrease. (After E. M. Purcell, *Electricity and Magnetism*, McGraw-Hill, New York, 1963.)

1–23 A particle of rest mass m, charge q, and initial velocity $\mathbf{v}_0$ enters a region of space containing a uniform electric field $\mathcal{E}$ perpendicular to $\mathbf{v}_0$. Find the subsequent trajectory of the particle, and show that the path is a parabola as long as the speed of the particle is much less than c. (After H. Goldstein, *Classical Mechanics*, Addison-Wesley, Reading, Mass., 1950.)

1–24 A uniform rod of mass M and length $2L$ spins with angular frequency $\omega \ll c/L$ about its center. Its angular momentum and kinetic energy are given by

$$l = \frac{ML^2\omega}{3}\left(1 + A\frac{\omega^2L^2}{c^2} + \cdots\right)$$

$$K = \frac{ML^2\omega^2}{6}\left(1 + B\frac{\omega^2L^2}{c^2} + \cdots\right)$$

What are the values of A and B?

1–25 The "classical radius" of the electron, r_0, is a combination of physical constants, numerically equal to 2.818×10^{-15} m. The electron is also known to have an intrinsic angular momentum, or spin, equal to $\hbar/2 = 5.272 \times 10^{-35}$ joule·sec. If the electron is assumed to be a uniform sphere of radius r_0 spinning with angular frequency ω, find the numerical value of ω. (Use a nonrelativistic analysis.) Do you think this is a reasonable model? Why not? The moment of inertia of a uniform sphere is $2MR^2/5$.

1–26 Calculate the relativistic increase of mass (in %) associated with the kinetic energy of an electron moving in the first Bohr orbit of (a) hydrogen ($Z = 1$), (b) uranium ($Z = 92$). The orbit is defined by $F = mv^2/r$ and by $mvr = \hbar = 1.05 \times 10^{-34}$ joule·sec. The force F is given by $F = kq_1q_2/r^2$, where $q_1 = Ze$, $q_2 = e = 1.6 \times 10^{-19}$ coulomb, and $k = 9 \times 10^9$ newton·m^2/coulomb^2.

Is there any point to which you would wish to draw my attention?
To the curious incident of the dog in the night-time.
The dog did nothing in the night-time.
That was the curious incident, remarked Sherlock Holmes.

THE MEMOIRS OF SHERLOCK HOLMES (1893)

The interpretation of these results is that there is no displacement of the interference bands.

A. A. MICHELSON (1881)

2

Perplexities in the propagation of light

THE EVIDENCE that we assembled in Chapter 1 leaves us in no doubt that a velocity equal to the velocity of light in vacuum has a deep significance in physics. Yet when we try to couple this fact with the classical dynamical scheme we seem to run into serious trouble. For at first glance a universal velocity is a contradiction in terms. We know that the velocity of a given motion has different measures in different frames of reference. How, then, is it possible to incorporate a unique velocity into the equations of dynamics? What justification do we have for speaking of *the* velocity of light, without reference to any particular frame of observation? (When we say, for example, that the speed of sound has a certain value, we know that this has meaning only as a statement of the speed of a wave with respect to the medium itself.) As we shall see in Chapter 3, it was Einstein's clear resolution of these questions, where others had made only tentative or partial attempts, that marked him as the true creator of special relativity. And a key factor in his success was his power to recognize (in the kind of way that Conan Doyle portrays Holmes as recognizing) the full significance of the things that did *not* happen. The most famous of all such phenomena is embodied in the results of what is universally known as the *Michelson-*

Morley experiment, but many other optical phenomena, some of them known long before the Michelson-Morley experiment was performed, contained clues to the solution of the problem. And in this chapter we shall review some of the important evidence relating to the propagation of light, both in empty space and in transparent material media.

THE NATURE OF LIGHT

The propagation of light involves the transport of energy away from a source. The simplest picture of this process is in terms of a stream of particles emitted from the source; Pythagoras, back in the 6th century B.C., proposed this mechanism. It accounts, very directly, for the propagation of light in straight lines (as evidenced by the sharpness of shadows) and for the fact that light can travel with complete ease through a vacuum. In 1667 there appeared the first clear exposition of a different theory—that light is a vibration communicated through a medium of some kind. This was propounded by Robert Hooke in his famous book *Micrographia*. At about this time were observed some of the phenomena that could not easily be related to a particle theory of light—the brilliant colors of thin air films between glass surfaces, and the encroachment of light upon the region of the geometrical shadow. Huygens, in his *Treatise on Light*,[1] developed the wave theory explicitly, and showed how it could account for reflection and refraction.

The particle theory and the wave theory have been the only clearly defined models by which to describe light and its propagation. For a long time—until the 20th century in fact—the two theories were taken to be mutually exclusive; it seemed obvious that acceptance of the one must imply rejection of the other. From the vantage point of today, we see that both photon and wave aspects of the behavior of light must be accepted—that the facts cannot all be forced into the mold of one or other of the two theories. We have learned also (thanks largely to Einstein) that we should focus on the bare facts of observation, and should not, through our adherence to a particular theory, read more into them than is there. To be specific, the wave properties of light are undeniable—diffraction, interference, polarization, etc.

[1]C. Huygens, *Treatise on Light* (written in 1678, published in 1690), unabridged republication of the original English edition of 1912 translated and introduced by S. P. Thompson, Dover, New York.

But the waves of ordinary experience require a medium. What more natural, therefore, than to build up a detailed specification of the medium that carries waves of light, and then to seek to detect it? Yet it was a quest that led only to frustration. Einstein showed that the search for the medium—the *luminiferous ether*—was sterile and unnecessary. The ether was a red herring—something that diverted physicists into following a false scent. Perhaps in this present discussion we should not introduce the ether at all, knowing that we are going to bury it again in the end. Yet one cannot fully appreciate the emergence of special relativity without some feeling for the importance and the appearance of reality that the ether once enjoyed. In the next section, therefore, we shall briefly discuss this background.

THE LUMINIFEROUS ETHER

The story of 19th-century physics was, in large part, the story of the triumph of the wave theory of light. At the beginning of the century (1801–1804) Thomas Young made his quantitative studies of interference phenomena. Beginning in 1818, Fresnel published calculations that were able to account in detail for the facts of interference, diffraction, and polarization. Since, as Huygens had shown, a wave theory was as competent as a particle theory to describe the ray properties of light—rectilinear propagation and the laws of reflection and refraction—the picture of light as a vibration in a medium, analogous to transverse waves on a string, seemed unassailable. But what could one say about the properties of the medium—which came to be called the *luminiferous ether*—in which these vibrations were presumed to take place?

Until about 1850 the propagation of light was envisaged in purely mechanical terms. This, however, posed very considerable difficulties, because it was hard to understand how the speed of light could be so very great. (The first numerical value of the speed of light—about 200,000 km/s—was obtained by Huygens in 1678 from measurements by the Danish astronomer Roemer in 1675. Roemer had noted systematic variations in the times, as recorded by clocks on earth, at which the moons of Jupiter moved into the planet's shadow, and was astute enough to recognize that these variations were linked to the position of the earth in its orbit and to the associated transit time of the light over a variable distance.) A wave speed of more than a hundred thousand miles per second was many orders of magnitude greater than the speed of any other mechanical disturbance, and demanded a medium which, although so tenuous that the planets could travel through it year after year with no detectable loss of speed, must nevertheless develop very strong restoring forces when displaced from equilibrium—since the speed of propagation of a wave depends on this restorative property of the medium. It was unsatisfying, too, that the only clue to the properties of the medium was the measured value of c itself; nothing was known a priori.

The situation was transformed when James Clerk Maxwell, in 1861, produced his electromagnetic theory of light. It now became possible to *predict* the numerical value of the speed of light for any given medium, in terms of measurable electric and magnetic properties of the medium. There was no longer such a gulf between ether and ordinary matter, although the intangibility of the ether might still seem mysterious. The wave theory seemed to have achieved its ultimate justification, and the ether a reality that could not be gainsaid.

Granted the existence of the ether, it was of course quite clear what was meant by "the speed of light." Any wave has a definite velocity with respect to the medium through which it moves. The magnitude of this velocity may be a function of wavelength (the phenomenon of dispersion) but is otherwise uniquely defined, at least for an isotropic medium (i.e., one containing no preferred directions). In particular, the speed of light through a medium should be quite independent of any motion of the source, in direct contrast to a particle-emission mechanism, in which one would expect the speed *relative to the source* to be the unique quantity. Acceptance of the wave theory did not wait upon an experimental proof that the measured value of c is indeed independent of the source velocity. If this had been known to Huygens, he would no doubt have used it as one more proof that a particle model of light was inadmissible. In fact, however, the wave theory appeared to be adequately supported by other lines of evidence, and the effect of source motion was not explored until the wave theory, in its turn, had run into severe difficulties. In Chapters 3 and 5 we shall have more to say about experiments on radiation from moving sources; for the present we shall merely state the result—that the velocity of a source of light is *not* communicated to the radiation it emits.

Let us, then, put ourselves in the position of a physicist of, say, 1900, and look at some striking optical phenomena from the standpoint of a wave theory.

In 1725 the British astronomer James Bradley tried to measure the distances of some stars by looking for an apparent change in their positions as the earth moved around the sun. He hoped to use the diameter of the earth's orbit as a base line, and to determine stellar distances in essentially the same way as a surveyor measures distances by triangulation. He did observe an effect, but he discovered that it was not parallax; it depended not on the earth's position, but on its *motion* at a given point in the orbit. (The true parallax effect is unobservably small for most stars.)

Consider Fig. 2–1, which depicts the orbit of the earth around the sun, and a star viewed from four positions of the earth, at 3-month intervals. The true altitude of the star with respect to the plane of the earth's orbit (the ecliptic) is the angle θ_0. Because of the earth's changes of position, one would expect the altitude to be greatest when the earth is at position 2 and

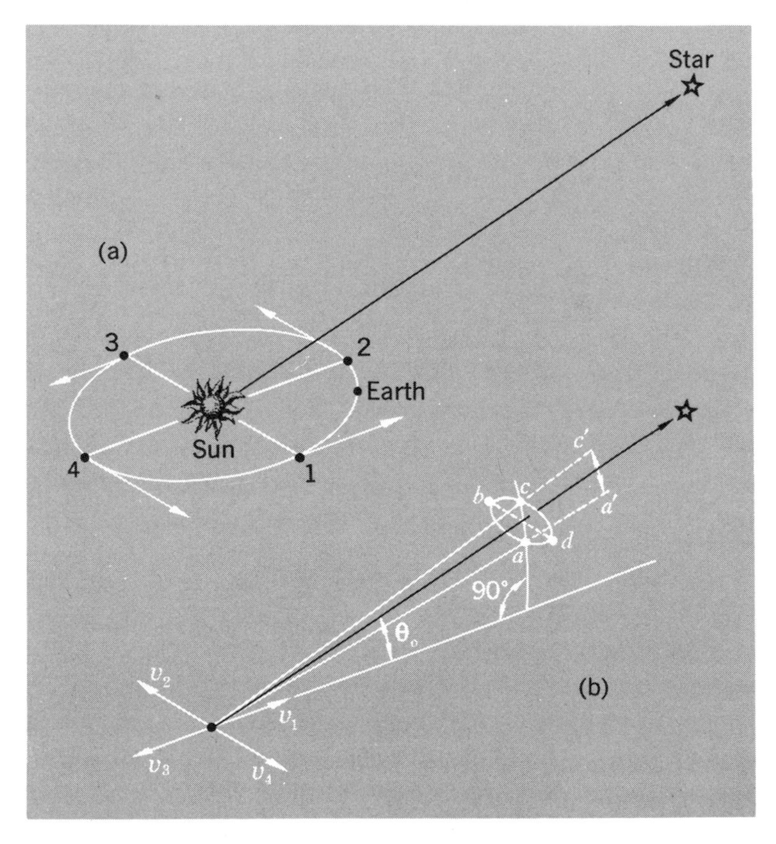

Fig. 2–1 Stellar aberration. (a) A distant star is viewed from the succession of positions 1-2-3-4 as the earth moves around the sun. (b) In a coordinate system attached to the earth (but with the direction of the axes fixed in space), the apparent position of the star follows the elliptical path a-b-c-d. The effect depends on the changes in the direction of the earth's velocity, not on the changes in its position as such.

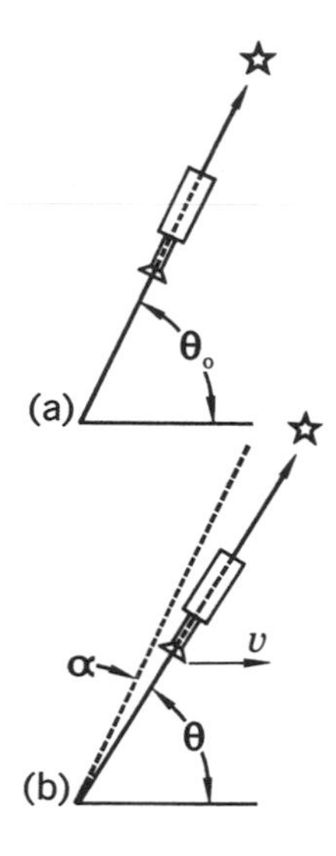

Fig. 2–2 Basis of stellar aberration. (a) A stationary telescope aligned on a star. (b) A moving telescope aligned on the same star.

least when it is at position 4. Instead, Bradley found that the altitude was greatest at position 3 and least at position 1.

The phenomenon can be understood in terms of Fig. 2–2. A telescope on a stationary earth (a) would have to be pointed at the true altitude θ_0 in order that the rays of light from the star should travel along the axis of the instrument and form an image at the center of the field of view. But on a moving earth (b) the telescope would have to be tilted at a slightly different angle, θ. The difference of angles is the aberration, α. We can observe a comparable phenomenon when it rains. If raindrops are falling vertically at speed w, but we are in a vehicle moving at speed v, we see the drops moving along straight lines inclined to the vertical at an angle $\tan^{-1}(v/w)$.

The aberration effect would never be detectable if the earth moved always with the same velocity, but the *changes* in the direction of motion during the year lead to a systematic change in the apparent position. This can be analyzed quantitatively with the help of Fig. 2–1. At positions 1 and 3 the earth's velocity vector and the line from sun to star make an angle θ_0 with one another. At positions 2 and 4 the earth's velocity is at right angles to the line from sun to star; the aberration angle has its greatest possible values ($\pm v/c$) at these positions. At positions 1 and 3 we have a situation like that depicted in Fig. 2–2, in which the aberration angle is only of magnitude $v \sin \theta_0/c$. Thus in the course of a year the star appears to describe an elliptical path which has a major axis (measured as an angle 2β) equal to $2v/c$ and a minor axis of $2\beta \sin \theta_0$. The length of the major axis should be the same for all stars; the length of the minor axis depends on the altitude θ_0 of a star with respect to the plane of the earth's orbit.

What Bradley observed corresponded exactly to the above description. Figure 2–3 is a graph of some of his observations on the star γ Draconis; it shows how the apparent position of the star varied in the north-south direction over a 12-month period.[1]

[1]Data taken from J. Bradley, *Phil. Trans. Roy. Soc.*, **35,** 637 (1729). For an interesting account of Bradley's work, with many details, see A. Stewart, "The Discovery of Stellar Aberration," *Sci. Am.* **210**(3), 100 (1964).

The east-west component of the aberration was not recorded. (You should consider the practical difficulties of measuring this component.) Thus Bradley's data span the minor axis of the aberration path.

Now the orbital speed of the earth is 30 km/sec, so that the value of $2v/c$ is 2×10^{-4} rad, or about 41 seconds of arc. A typical aberration path would resemble the outline of a football viewed from a distance of about 1 mile. The data shown in Fig. 2–3 span the minor axis of an aberration path for which $\theta_0 = 75°$, giving a calculated variation of 39.6″ between maximum and minimum altitudes, with a sinusoidal variation (why?) between these extremes. The observed range of variation corresponds extremely closely to this theoretical value. Actually, Bradley himself could not make a quantitative theoretical check of his result, because the speed of light was not well enough known. Instead, being sure that the basic interpretation of the phenomenon was correct, he used the observed aberration angles to obtain an improved value of c, the earth's orbital speed being at that time quite well known.

When we come to analyze the aberration phenomenon in terms of a theory of light, it is clear that a particle model provides a very ready explanation; it is just like the falling-rain analogy. However, one can also account for the effect in terms of waves

Fig. 2–3 Bradley's data on the north-south component of the aberration of γ-Draconis (1727–1728).

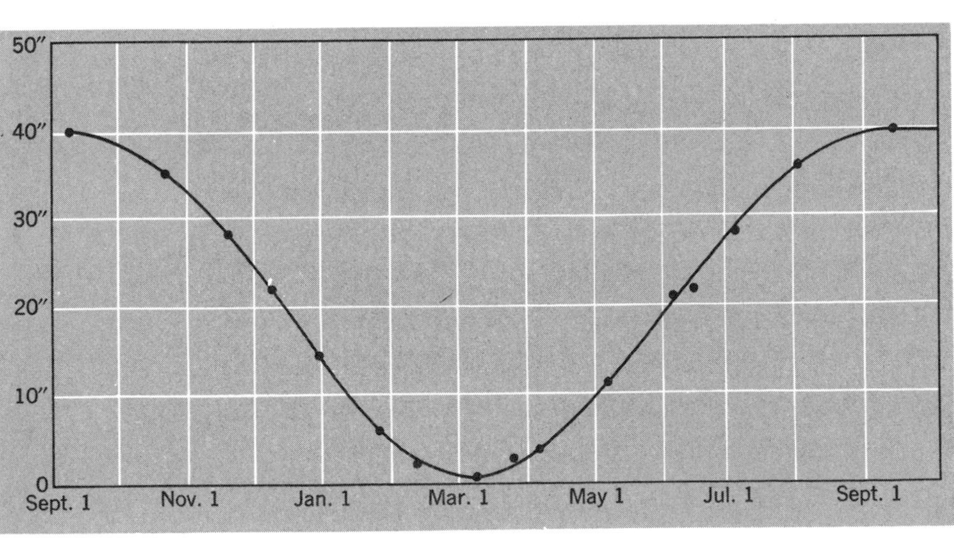

traveling through the ether, *provided* the ether remains completely undisturbed by the earth's motion. If, on the other hand, the ether near the earth were carried along with it, the aberration would not take place.[1] The notion of an ether completely undisturbed by the passage of the earth must have seemed a rather strained one to many physicists, but with the wave theory standing supreme it appeared unavoidable. And then it was natural to ask: Can one make any measurements that will disclose the magnitude of the velocity of the earth through the ether? We shall next describe some experiments bearing on this question.

A MODIFIED ABERRATION EXPERIMENT

Suppose that a telescope has been aimed at a star whose true direction is at 90° to the plane of the earth's orbit. Let the unknown aberration angle be α [Fig. 2–4(a)] and let the unknown speed of the earth through the ether be v. Now imagine that the whole tube of the telescope is filled with water, of refractive index n. Since light travels more slowly in water than in air or vacuum, the time for the light to travel down the length of the telescope tube will be lengthened—by the factor n. One might expect, therefore, that to keep the star's image in the center of the field of view one would have to tilt the telescope further, to some new aberration angle β, and that the amount of this adjustment could be used to find the speed v. At first glance one might think that the angle β would be just nv/c, but in analyzing this experiment one must remember that, because the objective lens of the telescope now has air on one side and water on the other, the light rays entering the telescope are bent toward the axis of the instrument, as indicated in Fig. 2–4(b). Inside the telescope we would expect the rays to travel at an angle δ to the axis such that

$$n = \frac{\sin \beta}{\sin \delta} \approx \frac{\beta}{\delta}$$

Since the light is traveling downward with speed c/n, and the telescope is moving sideways at speed v, the condition for centering the star's image in the telescope is

$$\delta \approx \frac{v}{c/n} = \frac{nv}{c} \tag{2–1}$$

[1]Actually, this conclusion is not inescapable, but one must postulate quite outlandish conditions to have "convected ether" *and* aberration.

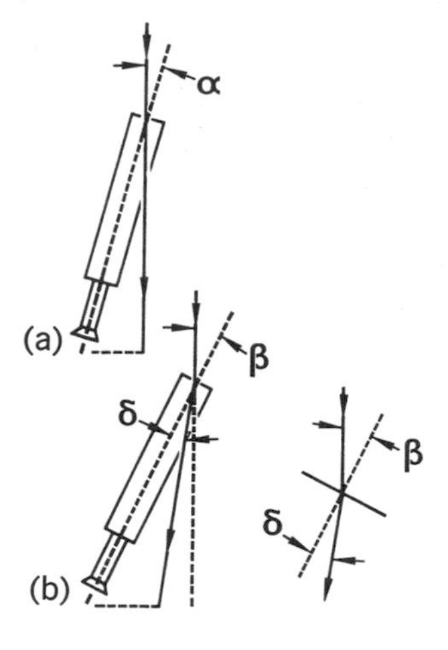

Fig. 2–4 Principle of Airy's experiment designed to reveal motion of the earth through the ether by sighting on a star with (a) a normal telescope; (b) the same telescope filled with water.

(Remember that the angles are grossly exaggerated in Fig. 2–4.) Now we do not know the true values of α, β, and δ, but we can surely measure the change of telescope direction, and we have

$$\beta \approx n\delta \approx \frac{n^2 v}{c} \qquad \alpha \approx \frac{v}{c}$$

Therefore,

$$\beta - \alpha \approx (n^2 - 1)v/c \tag{2–2}$$

Everything is directly measurable except v, the value of which we should therefore be able to discover. This very experiment was carried out by Sir George Airy in 1871. The result? There was absolutely no change in the apparent position of the star!

How can we explain this null result? As a matter of fact, it had been predicted by the brilliant J. A. Fresnel, who had suggested this experiment many years earlier. Fresnel's expectations were based, however, not on the fundamental impossibility of detecting absolute motion, but on the assumption of a partial drag of the light by the medium. He had postulated this in 1818, after his fellow-countryman Arago had found that the refraction of starlight through glass appeared to take place just as though the earth were at rest in the ether.

Sir George Airy's experiment can be easily analyzed in these terms. For suppose that the water drags the light sideways with a fraction f of its own velocity v. The experiment has shown that the angle β is equal to the original aberration angle α ($= v/c$) and hence that the angle δ is equal to α/n. Let the length of the telescope be l; then the time t for the light to pass down it when water-filled is nl/c. In time t the telescope moves through the distance vt; if the light is to emerge at the center of the eyepiece, its sideways displacement must be equal to this. But, as measured from the position of the eyepiece when the light enters the top of the telescope, the displacement of the light is the sum of $l\delta$, due to refraction, and fvt, due to dragging by the water. Hence we have

$$vt = l\delta + fvt$$

But

$$l = ct/n \qquad \text{and} \qquad \delta = \alpha/n = v/nc$$

Therefore,

$$vt = \frac{ct}{n}\frac{v}{nc} + fvt$$

whence

$$f = 1 - 1/n^2 \tag{2–3}$$

The quantity f is known as Fresnel's *drag coefficient.*

It may seem curious indeed that nature should provide a drag coefficient of just such a size that Airy's experiment, and others like it, should yield just the same result as if the earth were motionless with respect to the ether. Is there some way of exhibiting this drag as a *positive* effect, rather than as a null phenomenon? H. L. Fizeau had answered this in the affirmative in a famous experiment he performed in 1851.

FIZEAU'S MEASUREMENT OF THE DRAG COEFFICIENT

Fizeau set up the apparatus shown diagrammatically in Fig. 2–5. A beam of light from a source S falls on an inclined glass plate P that has a semitransparent metal coating such that the beam is split into two parts. One part travels straight on until it strikes a mirror M_1. The other part is reflected through 90° and strikes M_3. With a third mirror M_2 in place, the two beams travel around the same rectangular path but in opposite directions. When they arrive back at P, part of the first beam is reflected and part of the second is transmitted, and the light thus emerging from the system enters a telescope T.

This arrangement constitutes a type of optical interferom-

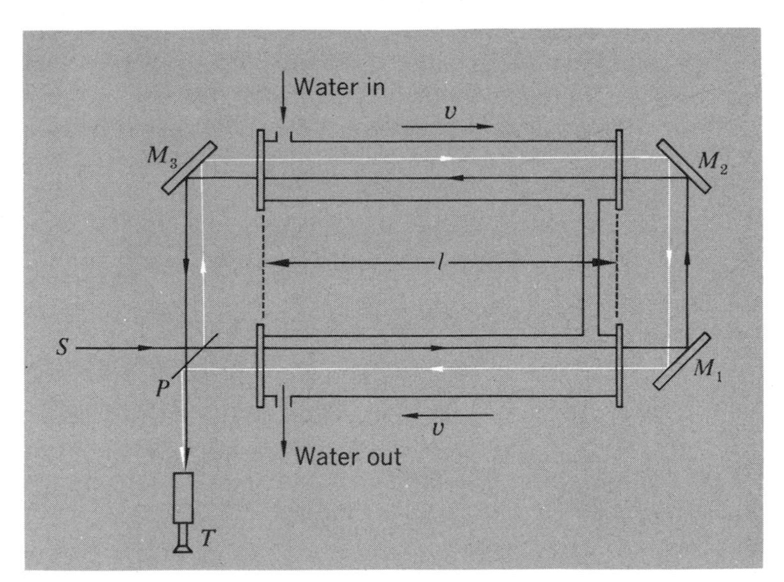

Fig. 2–5 Schematic diagram of Fizeau's "ether-drag" apparatus.

eter. If monochromatic or nearly monochromatic light is used, interference fringes are seen when one looks through the telescope. A particular fringe represents a particular optical path difference between the two interfering beams. (By *optical path* we mean the distance in vacuum equivalent to any actual path. A distance d through a medium of refractive index n represents an optical path nd; it is this that defines the number of wavelengths of the light that can be fitted into the distance.) The view through the telescope is like that shown in Fig. 2–6.

To provide a dragging effect, water is made to flow through two tubes with flat glass end plates as shown, so that one beam of light always travels with the water and the other beam always against it. Outside the water tubes the conditions are the same for both beams; thus to compute the optical path difference we need only consider what goes on inside the tubes. We can calculate this difference in terms of the difference of *times* for the two beams. If each tube is of length l and the speed of the water is v (with drag coefficient f), we have

$$\Delta t = \frac{2l}{(c/n) - fv} - \frac{2l}{(c/n) + fv}$$

which gives

$$\Delta t \approx \frac{4n^2 fvl}{c^2} \qquad (2\text{–}4)$$

Fig. 2-6 Interference fringes in apparatus of the Fizeau type. (Photo courtesy of G. C. Babcock, Michelson Laboratory China Lake, Calif.)

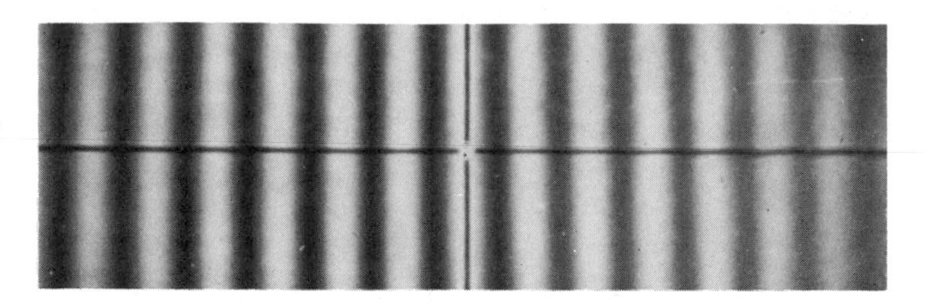

This implies an optical path difference $c\,\Delta t$. The change of optical path, expressed as a multiple (δ) of the wavelength λ of the light, is thus given by $c\,\Delta t/\lambda$:

$$\delta = \frac{4n^2 fvl}{\lambda c} \tag{2-5}$$

In Fizeau's experiment the approximate values were

$$l = 1.5\text{ m}$$
$$v = 7\text{ m/sec}$$
$$\lambda = 5.3 \times 10^{-7}\text{ m}$$
$$n = 1.33\text{ (refractive index of water)}$$
$$\delta = 0.23\text{ fringe}$$

Substituting these values in Eq. (2–5) gives the observed value of f:

$$f_{\text{obs}} \approx 0.48$$

The value of f calculated from Eq. (2–3) is

$$f_{\text{calc}} = 0.43$$

This could be taken as confirmation of the drag hypothesis. The experiment was, however, repeated with greater precision by Michelson and Morley in 1886, and still later became the subject of a series of beautiful investigations by P. Zeeman and his associates in Holland during the years 1914–1922.

The result of Fizeau's experiment could be taken as reinforcing the observations on stellar aberration. Both could be interpreted by supposing that a moving object does not communicate any of its motion to the ether, either outside or inside it. Inside a moving transparent material (according to this view) the light is carried partly by the material and partly by the ether that permeates it. Since the ether remains at rest, the light behaves as if only a fraction of the velocity of the material were added to the light. The question therefore remained: Could the motion of the earth through the ether somehow be detected?

PRELUDE TO THE MICHELSON-MORLEY EXPERIMENT

In 1879 Clerk Maxwell, in England, wrote an acknowledgment of some astronomical tables he had received from D.P. Todd of the U.S. Nautical Almanac Office in Washington. These tables contained many observations of the planet Jupiter. Maxwell, in his letter, asked about the possibility of measuring the velocity of the solar system through the ether by observing the eclipses of Jupiter's moons. (We have mentioned earlier how Huygens deduced the speed of light from Roemer's data on the time lag in detecting these eclipses.)

The essence of Maxwell's idea was very simple. Jupiter has a period of 12 terrestrial years, and so in half a terrestrial year, while the earth moves from A to B (Fig. 2–7), Jupiter does not travel very far in its orbit. Thus by observing the apparent times of eclipses with the earth successively at A and at B, we can infer the time taken for light to travel a distance equal to the diameter of the earth's orbit. This was Roemer's discovery, in fact. But if this time is measured when Jupiter is first at A', and then, 6 years or so later, at B', we can hope to discover whether the whole solar system is moving through the luminiferous ether with

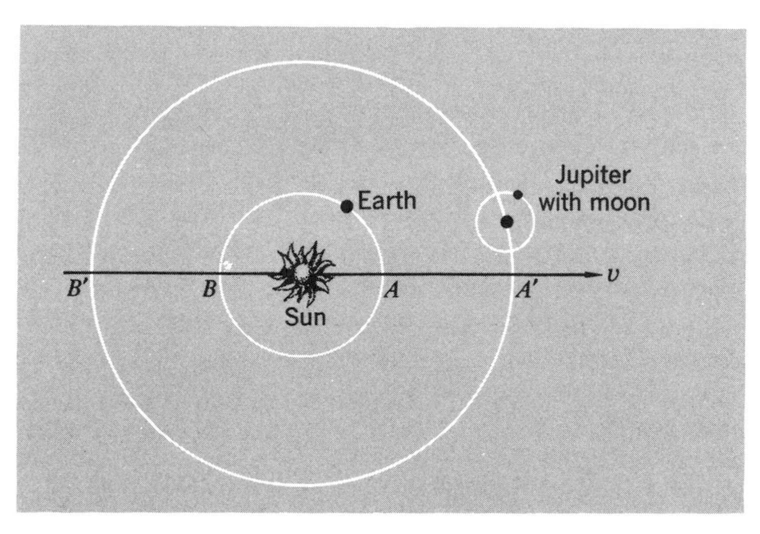

Fig. 2–7 Orbits of earth, Jupiter, and one of Jupiter's moons. Intervals betweeen moon's eclipses behind Jupiter, as observed at earth, depend on relative positions and motions of earth and Jupiter.

some speed v. For, if the diameter of the earth's orbit is l, we would expect to have

$$t_1 = \frac{l}{c + v} \qquad t_2 = \frac{l}{c - v}$$

and hence a time difference Δt given by

$$\Delta t = t_2 - t_1 \approx \frac{2lv}{c^2} = \frac{2v}{c} t_0 \tag{2–6}$$

where $t_0 = 16$ min approximately. If we could detect $\Delta t = 1$ sec, this would then correspond to v equal to about 150 km/sec. This is rather high compared with the known velocities of stars relative to the solar system (20 km/sec is a typical figure), but it is not excessive. Unfortunately, however, the difficulty of establishing any such difference through measurements made 6 years apart is great, and in fact the astronomical data available to Maxwell were not accurate enough for any such analysis, as Todd pointed out in his reply.

Maxwell, in proposing the above method, pointed to the feature that it was a first-order experiment—the effect would be proportional to the first power of the ratio v/c. And in his letter to Todd, Maxwell remarked that this distinguished it from terrestrial experiments on the speed of light, because these experiments necessarily used a beam of light that returned to its starting point. The time for any such round trip does in principle depend on the speed of the earth through the ether, but the effect is of the second order. Thus, if the length of the path (one way) is l, and if the earth's motion happens to be along the direction of the path at speed v, the total time taken by the light would be given by

$$t = \frac{l}{c + v} + \frac{l}{c - v} = \frac{2lc}{c^2 - v^2} \approx \frac{2l}{c}\left(1 + \frac{v^2}{c^2}\right)$$

The change of time due to the motion is thus given by

$$\Delta t \approx \frac{2l}{c} \frac{v^2}{c^2} \tag{2–7}$$

Maxwell remarked that this effect would be undetectably small. If v were taken as the orbital speed of the earth, we should have $v/c = 10^{-4}$, so that the fractional variation of flight time would be only 1 part in 10^8, which would surely be beyond the limits of observation. But Maxwell's letter was read by A. A. Michelson, who in the previous year (1878) at the young age of

25 had carried out a superb measurement of the speed of light.[1] And Michelson did not accept without question the impossibility of detecting motion via Eq. (2–7). Instead, he began thinking about a method to achieve it. Two years later, in 1881, he had some results. The next section describes his experiment, the most famous of all attempts to detect our motion through absolute space as defined by the ether.

THE MICHELSON-MORLEY EXPERIMENT

Michelson, who became the first American to win a Nobel prize in science (it was awarded to him in 1907), invented a new instrument of unprecedented sensitivity to look for the effect that Maxwell had discounted. The essential features of his apparatus, known universally as the Michelson interferometer, are shown in Fig. 2–8. Light from a source S falls on an inclined glass plate P that has a semitransparent metal coating on its front face. This splits the light into two parts. One part travels on through the plate and strikes a mirror M_1. It retraces its path to the point where the beam was first split, and a fraction of it is reflected back through the plate into a telescope T. A second light path is by reflection from the beam splitter to a mirror M_2 and back.

[1]Michelson, a young naval instructor, had just been transferred to the Nautical Almanac Office where Maxwell's correspondent Todd worked.

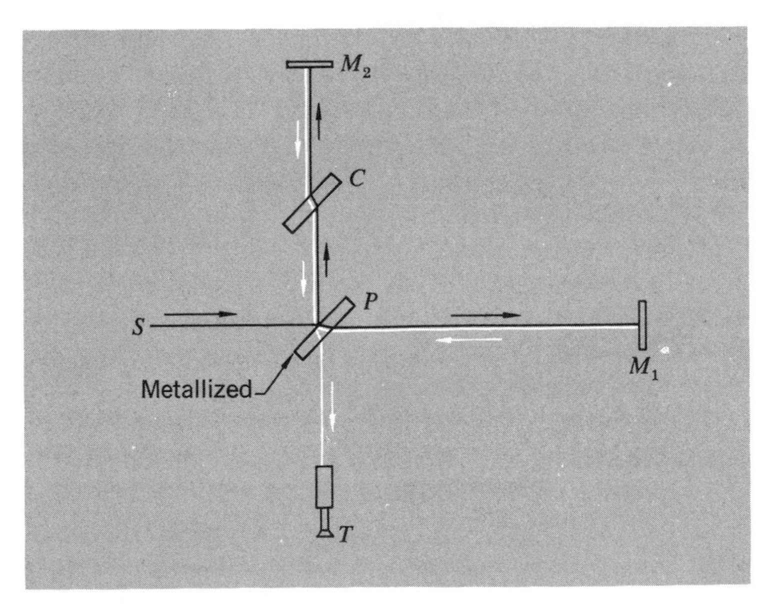

Fig. 2–8 Schematic arrangement of Michelson interferometer.

The Michelson-Morley experiment

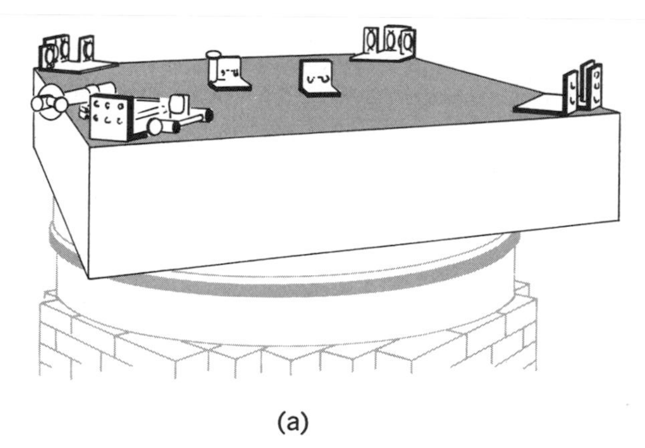

(a)

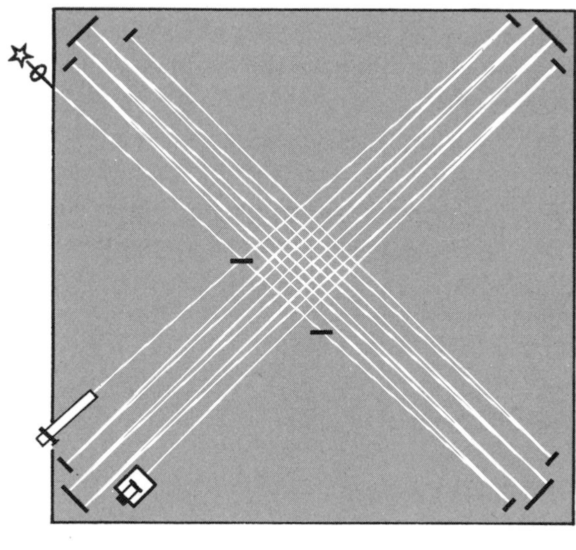

(b)

(a) A sketch of the apparatus. (b) Plan view of the optical system. (c) Variation of fringe position during one rotation of the apparatus. [Diagrams (a), (b), and (c) are adapted from A. A. Michelson, Studies in Optics, Univ. Chicago, Phoenix Books, 1962.]

The photograph shows fringes during two complete rotations of the apparatus in a later repeat of the experiment by G. Joos in 1930. *(From Joos, Lehrbuch der Theoretischen Physik, Akademische Verlagsgesellschaft, Leipzig.) The table summarizes the results of various studies of the Michelson-Morley experiment.* (δ_{calc} = *Fringe shift according to Eq.* (2–14), *for* v = 30*km/sec.*)

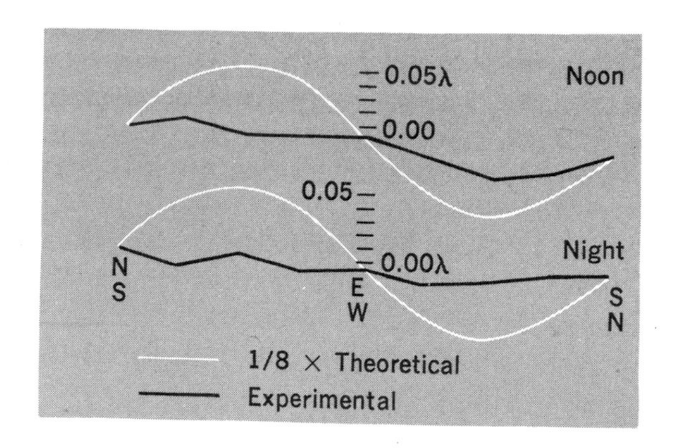

(c)

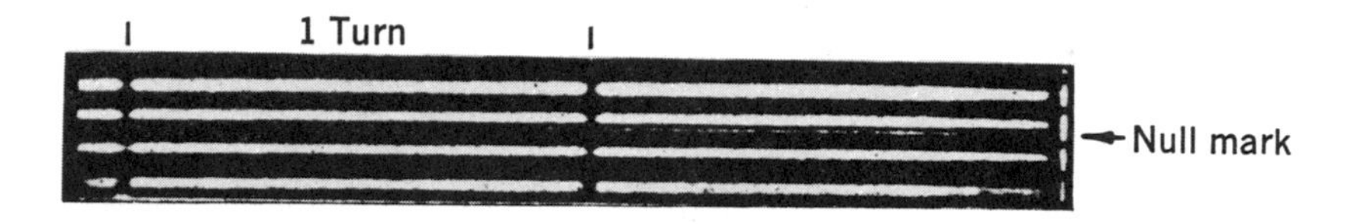

TRIALS OF THE MICHELSON-MORLEY EXPERIMENT[a]

Observer; year	*l, cm.*	δ_{calc}	δ_{obs} *(upper limit)*	*Ratio*
Michelson; 1881	120	0.04	0.02	2
Michelson and Morley; 1887	1100	0.40	0.01	40
Morley and Miller; 1902–1904	3220	1.13	0.015	80
Miller, 1921	3220	1.12	0.08	15
Miller; 1923–1924	3220	1.12	0.03	40
Miller (sunlight); 1924	3220	1.12	0.014	80
Tomaschek (starlight); 1924	860	0.3	0.02	15
Miller; 1925–1926	3200	1.12	0.08	13
Kennedy; 1926	200	0.07	0.002	35
Illingworth; 1927	200	0.07	0.0004	175
Piccard and Stahel; 1927	280	0.13	0.006	20
Michelson et al; 1929	2590	0.9	0.01	90
Joos; 1930	2100	0.75	0.002	375

[a]From a review by Shankland et al., *Rev. Mod. Phys.*, **27**, 167 (1955).

Fig. 2–9 Straight fringes formed by Michelson interferometer. (Photo by Jon Rosenfeld, Education Research Center, M.I.T.)

A compensating plate C causes this beam to travel through the same thickness of glass as the first one (in the interests of optical symmetry) before it rejoins the first beam and passes with it into the telescope. If monochromatic light is used, interference fringes are seen when one looks through the telescope.

If the plate P is inclined at 45° and the surfaces of the mirrors are almost but not quite at 90° to each other, one obtains fringes resembling those formed by a wedge of very small angle. By suitable adjustment of the mirrors these fringes can be made horizontal. The view through the telescope is then as indicated in Fig. 2–9. If the optical path lengths (PM_1 and PM_2) giving rise to a particular fringe are l_1 and l_2, we have the condition

$$2(l_1 - l_2) = m\lambda \tag{2–8}$$

where m is some integer. If l_1 or l_2 is changed by $\lambda/2$, e.g., by moving one of the mirrors, the pattern of Fig. 2–9 moves by one fringe interval over the telescope cross-hair XX′.

Let us now imagine that Michelson's whole apparatus is moving at speed v in the direction PM_1, with respect to the inertial frame defined by the hypothesized ether. From the viewpoint of the laboratory, an "ether wind" is blowing past the apparatus (Fig. 2–10). The light traveling from P to M_2 and back must be aimed into the wind at such an angle that the resultant velocity is along PM_2. It would follow from the usual Galilean law of composition of velocities that the magnitude of the resultant velocity relative to the interferometer is $(c^2 - v^2)^{1/2}$. The light traveling between P and M_1 would have (again relative to the interferometer) a resultant velocity $c - v$ when traveling toward M_1 and a resultant velocity $c + v$ on the return trip.

We can thus calculate the times taken for light to travel from P to the mirrors and back:

$$t_1 = \frac{l_1}{c - v} + \frac{l_1}{c + v}$$

i.e.,

$$t_1 = \frac{2l_1c}{c^2 - v^2} = \frac{2l_1/c}{(1 - v^2/c^2)} \tag{2–9}$$

$$t_2 = \frac{2l_2}{(c^2 - v^2)^{1/2}} = \frac{2l_2/c}{(1 - v^2/c^2)^{1/2}} \tag{2–10}$$

This defines a time difference Δ, which for $v \ll c$ is given approximately as follows:

$$\Delta = t_1 - t_2 \approx \frac{2l_1}{c}\left(1 + \frac{v^2}{c^2}\right) - \frac{2l_2}{c}\left(1 + \frac{v^2}{2c^2}\right)$$

i.e.,

$$\Delta \approx \frac{2(l_1 - l_2)}{c} + \frac{2l_1v^2}{c^3} - \frac{l_2v^2}{c^3} \tag{2–11}$$

If now the whole apparatus is turned through 90°, so that PM_2 now points along the direction of motion, we have a new time difference Δ':

$$\Delta' = t_1' - t_2' \approx \frac{2l_1}{c}\left(1 + \frac{v^2}{2c^2}\right) - \frac{2l_2}{c}\left(1 + \frac{v^2}{c^2}\right)$$

i.e.,

$$\Delta' \approx \frac{2(l_1 - l_2)}{c} + \frac{l_1v^2}{c^3} - \frac{2l_2v^2}{c^3} \tag{2–12}$$

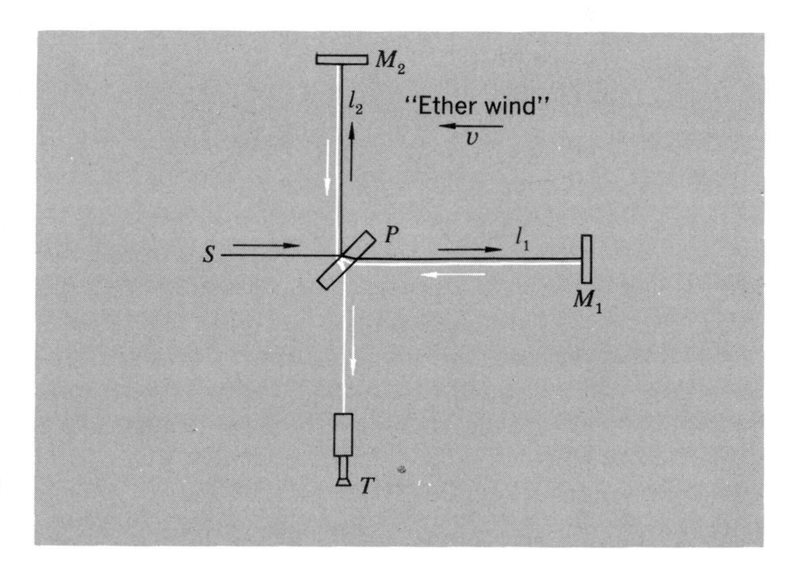

Fig. 2–10 Illustrating principle of Michelson-Morley experiment in terms of "ether wind."

The change of time difference would lead to a shift of the interference pattern by an amount corresponding to δ fringes, where $\delta = c(\Delta - \Delta')/\lambda$, i.e.,

$$\delta = \frac{(l_1 + l_2)v^2}{\lambda c^2} \tag{2-13}$$

If $l_1 = l_2 = l$, we can express this result in the following form:

$$\delta = \frac{2(v/c)^2}{\lambda/l} \tag{2-14}$$

The values of λ, l, and c are known, but what should we put for v? It seemed to Michelson and to everyone else concerned with the problem that one clearly identifiable contribution was the velocity of the earth in its orbit—about 30 km/sec. This, as we have seen, would give $v/c \approx 10^{-4}$. We can put $\lambda \approx 6 \times 10^{-7}$ m, and in Michelson's first apparatus $l = 1.2$ m, so that $\lambda/l \approx 5 \times 10^{-7}$. Putting these together gives $\delta \approx 0.04$ fringe. This is a small effect but in Michelson's skilled hands would have been measurable. Yet to Michelson's undoubted surprise and disappointment, when he set up his interferometer and rotated it through 360° there was no significant shift of the fringe pattern. His terse statement of this result, quoted from his first paper on the subject,[1] is at the beginning of this chapter. He added the comment: "The result of the hypothesis of a stationary ether is thus shown to be incorrect."

The null result was so unexpected, and so hard to account for, that strenuous efforts were made, especially by the great Dutch physicist H. A. Lorentz, to reconcile it with other optical phenomena such as we have described in this chapter. Lorentz devised a theory according to which, without contradicting the stellar-aberration experiments, one might have a partial drag of the ether by the earth. A more precise repetition of Michelson's experiment became desirable, and Lord Rayleigh wrote to Michelson urging him to do it. Thus it came about that Michelson, now in collaboration with E. W. Morley, undertook a much more precise investigation, based upon optical paths about 10 times longer than in the first experiment. This made the expected fringe shift about 0.4 fringe, but the observed effect was at most 0.005 fringe. It is this refined version of the experiment, performed in 1887,[2] that has long been regarded as one of the main

[1]A. A. Michelson, *Am. J. Sci*, **122,** 120 (1881).

[2]A. A. Michelson and E. W. Morley, *Am. J. Sci.*, **134,** 333 (1887).

experimental pillars of special relativity. See pages 52 and 53 for a diagram of the Michelson-Morley apparatus and a sample of the results.

CONCLUDING REMARKS

The phenomena we have described in this chapter are such as to cause a strong sense of frustration in anyone who wishes to hold to the picture of light as waves in the ether. Every experiment designed to reveal our motion through this medium has the same result as though no motion existed or as though the medium itself did not exist. The result of the Michelson-Morley experiment, like stellar aberration, is readily explained in terms of a ballistic model of light, in which the speed is uniquely defined with respect to the source, not with respect to a medium. It would be in perfect accord with what Galileo first recognized—that motions on a uniformly moving ship or other inertial reference frame do not, as observed within that frame, reveal any information about the velocity with which the whole system is translated. But in the view of most physicists at the end of the 19th century a particle model was not relevant; one had to seek the explanations in the context of what seemed to be clearly a problem in wave propagation.[1] And looking at the evidence, one could see that the only observations which did not give null results were those in which well-defined *relative* velocities were involved. In stellar aberration it was the velocity of the earth at some point in its orbit relative to the velocity at some other point. In the Fizeau experiment it was the velocity of the flowing water relative to the rest of the apparatus. And of particular concern was the seemingly direct conflict, from the standpoint of a wave theory, between the existence of stellar aberration and the absence of any fringe shift in the Michelson-Morley experiment.

By way of summary to the discussions and presentation of evidence in this chapter, we show in Table 2–1 the way in which the various observational results fit into a description of light as particles traveling through space or as waves propagating through a luminiferous medium. This may help to give a better idea of how the different pieces of evidence bear on the general problem of the propagation of light.

[1]Perhaps the most damning evidence against any particle theory was the experimental proof (first obtained by Foucault and Fizeau in 1850) that light travels more slowly in water than in air, whereas the facts of refraction would, in terms of a particle model, require just the opposite to be true.

TABLE 2-1: EVIDENCE BEARING ON THE NATURE OF LIGHT

	Particle model	*Wave/ether model*
1. Light travels in straight lines	✓	OK if wavelength $\ll$ beam width
2. Interference and diffraction effects	No convincing explanation	✓
3. Polarization of light	No convincing explanation	✓
4. Light velocity independent of source velocity	*Definite disagreement*	✓
5. Speed of light greater in air than in water	*Definite disagreement*	✓
6. Fizeau experiment and Airy (water-filled telescope) experiment	Requires partial drag of light by medium	Requires partial drag of light by medium
7. Stellar aberration (Bradley)	✓	OK if earth moves with respect to ether
8. Michelson-Morley experiment	✓	Implies that earth does *not* move with respect to ether

If one regards the table as a kind of score-sheet, one sees that each theoretical model wins several checkmarks, corresponding to phenomena that are readily and convincingly explained in terms of the model in question. The particle model appears, however, to be definitely disqualified by the observations that the velocity of light is completely unaffected by the motion of its source, and that light travels faster in air than in water (or other dense media). Turning then to the wave theory, which indeed has been the main basis of our discussions throughout this chapter, we are confronted by the seeming conflict and incompatibility between the last two results listed in the table—stellar aberration and the Michelson-Morley experiment. Einstein was the first person to realize that the situation could not be met by a mere patching up of traditional theories. His radical and profound solution to the problem is the chief subject of Chapter 3.

PROBLEMS

2-1 (a) The head of SPECTRE has devised a fiendish plot to liquidate his mortal enemy 007; he will cause a toxic rain to fall on the deserted valley where the intelligence agent is staying with a companion. Fortunately Mr. Bond's latest bag of tricks contains an umbrella. If the rain falls at a speed of 10 m/sec, at what angle with the vertical must 007 hold the umbrella to stay dry as he and his companion run at a speed of 10 miles/hr to catch the next bus out of the deserted valley?

(b) A star in the sky is observed from earth to describe an elliptical path whose minor axis subtends an angle of 36″. What angle does the star make with the ecliptic?

(c) Suppose the apparent elliptical motion in part (b) were due entirely to the parallax effect. Approximately how far away (in light-years) from the sun would the star be?

2–2 The following is a sample of the data from the Fizeau "ether-drag" experiment as repeated in 1886 by Michelson and Morley:

Mean wavelength of light used	5700 Å
Length of each tube	6.15 m
Velocity of water flow	7.65 m/sec
Mean fringe shift upon reversal of flow	0.86 ± 0.01

Compare the value of the "drag coefficient" implied by these data with the value of $1 - 1/n^2$ for water ($n = 1.33$).

2–3 The Fizeau experiment as described in the text requires a rapid flow of water through the apparatus, often leading to turbulence. A possible alternative to this situation would be to pass light through a rotating disk of optical glass (see the figure).

(a) Show that the optical path difference between the beams is

$$2n^2 fl\omega(4R^2 - l^2)^{1/2}/c$$

where n is the index of refraction of glass, f the drag coefficient, ω the angular frequency of rotation, and c the speed of light.

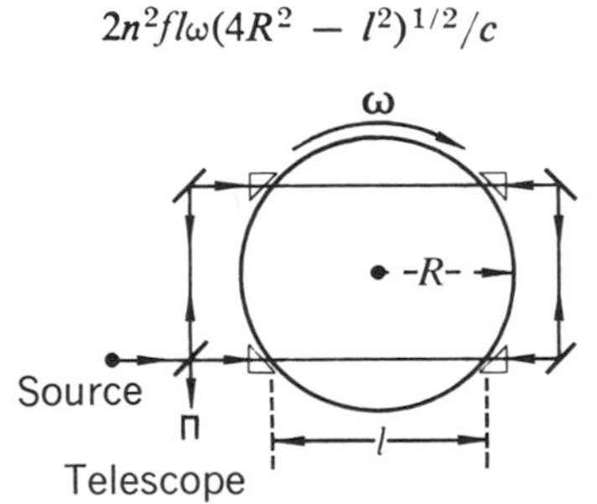

(b) If $f = 1 - n^{-2}$, $n = 1.50$, $l = 20$ cm, $R = 20$ cm, and $\lambda = 5300$ Å, how fast must the disk be rotated to produce a maximum optical path difference equal to 0.2λ?

2–4 In a modern Fizeau experiment, a laser is in one arm of a triangular arrangement of mirrors (see the figure). A slab of glass ($n = 1.5$) 1 cm thick is inserted in another arm of the system. The laser light can travel in a closed path in either direction. When the slab is stationary the two optical paths are identical. Suppose now that the slab is moved with a speed of 1 cm/sec in the direction indicated. What is now the optical path difference? (This difference, although minute, leads to a splitting of the laser frequency which can be converted to an audible beat note if samples of the oppositely circulating beams are suitably combined in a detector.) Note: This

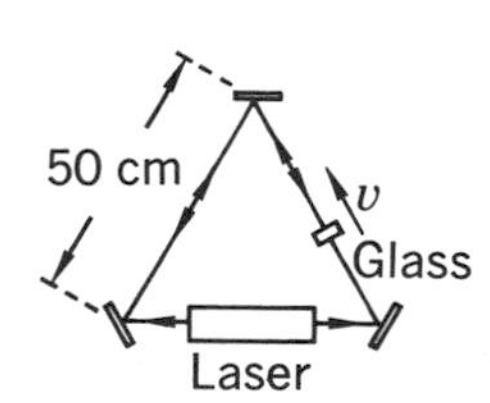

differs importantly from the original Fizeau arrangement, in that the boundaries of the medium change position while the light passes through. Your calculation must take account of this fact.

2–5 The text refers to Huygens' determination of the speed of light from Roemer's study of the apparent variations in the time intervals between successive eclipses of Jupiter's moons. To analyze this quantitatively, consider the positions P and Q of the earth corresponding to observations of two successive eclipses of one of Jupiter's moons (see the figure).

(a) Show that, if the true time interval between eclipses is τ (equal to the orbital period of a moon), the observed time interval is $\tau + \Delta\tau$, where $\Delta\tau \approx R\,\Delta\theta \sin\theta/c$ and $\Delta\theta \approx v\tau/R$. Hence show that the accumulated time lag as the earth moves from A to B (ignoring Jupiter's own change of position) is $2R/c$.

(b) The period of Jupiter's second moon, Europa, is 3.55 days. What is the maximum discrepancy between this true period and the time interval between successive eclipses as observed at the earth? What is the accumulated time lag between A and B? (Earth's orbital speed = 30 km/sec; orbital radius = 1.49×10^8 km.)

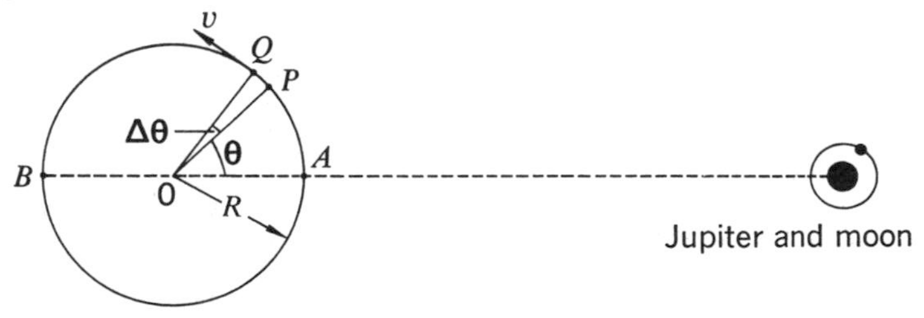

2–6 In one arm of a Michelson interferometer there is placed a closed tube of length 0.2 m with transparent end pieces through which the light passes freely. A fringe pattern is observed using light of wavelength 5.9×10^{-7} m, and then the air is evacuated out of the tube. By how many fringes will the fringe pattern shift? The speed of light in air is $(1 - 2.9 \times 10^{-4})c$.

2–7 In the Michelson-Morley experiment of 1887, the length l of each arm of the interferometer was 11 m, and sodium light of wavelength 5.9×10^{-7} m was used. The experiment would have revealed any fringe shift larger than 0.005 fringe. What upper limit does this place on the speed of the earth through the supposed ether?

2–8 The ether-wind theory of the Michelson-Morley experiment is discussed in the text for the special case where the arms of the interferometer are parallel and perpendicular to the wind. Consider the general case for an angular setting θ as shown (see the figure). Prove that, for equal arms of length l, the time difference for the two paths is given to a good approximation by

$$\Delta t(\theta) = \frac{v^2 l}{c^3} \cos 2\theta$$

2–9 A Michelson interferometer for sound waves is made with a loudspeaker as source, a microphone as detector, a thin sheet of paper as beam splitter, and two pieces of board as mirrors (see the figure). The arms of the interferometer are of equal length l, and the whole apparatus can be rotated. A wind, constant in speed and direction, blows past the apparatus. Once every second the speaker emits a very brief pulse of sound. Answer the following questions, using the result of Problem 2–8.

(a) In most orientations of the interferometer the echo received by the microphone consists of two pulses in quick succession. Why?

(b) At certain orientations of the interferometer the time delay between the two echoes is maximum. How many such orientations are there in one complete revolution of the apparatus? In relation to the direction of the wind, what are these orientations?

(c) If the length of each arm of the interferometer is 3 m, and the maximum time delay between echoes is found to be 10^{-5} sec, what is the speed of the wind? Sound waves travel with a constant velocity relative to the air. Take the magnitude of this velocity to be 300 m/sec.

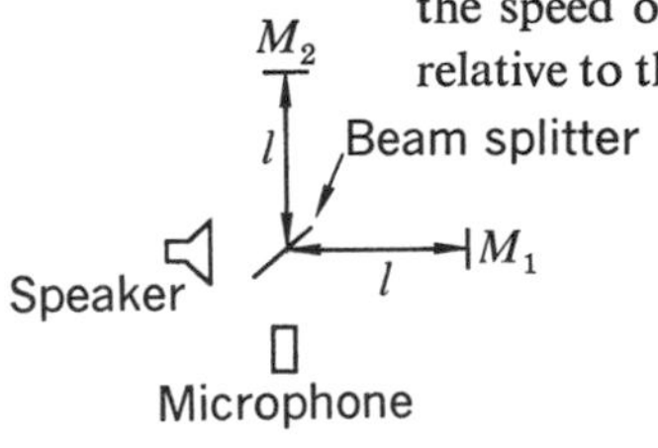

In the autumn of [1905] . . . *Einstein published a paper which set forth the relativity theory of Poincaré and Lorentz with some amplifications, and which attracted much attention.*

E. T. WHITTAKER,

History of the Theories of Aether and Electricity (1953)

This quotation has become famous for its failure to give adequate credit to Einstein. See the discussion in the Epilogue of this book.

3

Einstein and the Lorentz-Einstein transformations

PREAMBLE: THE CONTRACTION HYPOTHESIS

WHY DID the Michelson-Morley experiment yield a null result? It was a question that troubled some of the finest minds in 19th-century physics. There was no doubt about the facts; as we saw in Chapter 2, they were supported by a number of other optical observations which all told the same story—that nature would apparently yield up no information about our state of motion with respect to a supposed fundamental frame of reference. And this view of things was precisely the one that most physicists took: the ether existed, and motion through it was real, but compensating effects were at work. The most famous of such special mechanisms was that devised by the Irishman G. F. FitzGerald and the Dutchman H. A. Lorentz (independently) in 1892, to explain the result of the Michelson-Morley experiment—a real contraction of a body along its direction of motion through the ether. If the contraction is by the factor $(1 - v^2/c^2)^{1/2}$, then a zero fringe shift follows directly. This is evident almost by inspection if one considers the situation (as we did in Chapter 2) in which the arms of the Michelson interferometer are assumed to lie along and perpendicular to the direction of the earth's motion. The times for the light to travel along its two possible paths are as already derived [Eqs. (2–9) and (2–10)]:

$$t_1 = \frac{2l_1/c}{1 - v^2/c^2}$$

$$t_2 = \frac{2l_2/c}{(1 - v^2/c^2)^{1/2}} \tag{3-1}$$

Suppose that a rod of normal length l_0 preserves that length if moving transversely to itself but shrinks to $l_0(1 - v^2/c^2)^{1/2}$ if moving parallel to itself. Then in Fig. 3–1(a) we should have

$$l_1 = l_{10}(1 - v^2/c^2)^{1/2} \qquad l_2 = l_{20}$$

Hence we should have

$$t_1 = \frac{2l_{10}/c}{(1 - v^2/c^2)^{1/2}} \qquad t_2 = \frac{2l_{20}/c}{(1 - v^2/c^2)^{1/2}}$$

and so

$$t_1 - t_2 = \frac{2(l_{10} - l_{20})/c}{(1 - v^2/c^2)^{1/2}} \tag{3-2}$$

Thus, if the interferometer were turned through 90°, so that the two arms exchanged roles, the time difference would again have the same value. These two positions are obviously rather special ones, but further calculation shows that in an arbitrary orientation, as in Fig. 3–1(b), the time difference is still correctly given by Eq. (3–2) if one assumes that, for an inclined arm, the component of length parallel to **v** shrinks and the component transverse to **v** remains unchanged. Thus with arm 1, for example, the relation between l_1 and the normal length l_{10} would be given by

$$l_{10}^2 = (l_1 \sin\theta)^2 + \frac{(l_1 \cos\theta)^2}{1 - v^2/c^2} \tag{3-3}$$

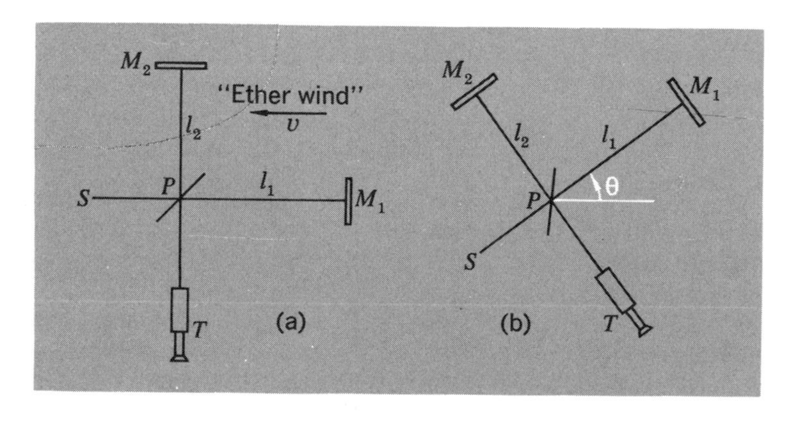

Fig. 3–1 (a) A Michelson-Morley apparatus with one arm headed directly into a supposed "ether wind." (b) The same apparatus in an arbitrary orientation.

You should satisfy yourself that this condition, and an analogous one for arm 2, lead to Eq. (3–2) for any value of θ.

EINSTEIN REASSERTS RELATIVITY

To this situation of clever (indeed, very distinguished) but essentially ad hoc theorizing, Einstein brought a piercing clarity of outlook—the kind of exalted simplicity that belongs to genius. Instead of imposing preconceived ideas on the facts, he let himself be guided by a bare minimum of data to a fundamental revision of ideas. It was an astonishing achievement in a field full of intellectual giants. At the beginning of his wonderful paper in which special relativity was brought into existence,[1] he comments on the fact that in such phenomena as the mutual interaction of a magnet and a conductor, it is only the relative motion that matters, and not the separate motion of either. And he goes on as follows:

> Examples of this sort, together with the unsuccessful attempts to discover any motion of the earth relatively to the "light medium," suggest that the phenomena of electrodynamics as well as of mechanics possess no properties corresponding to the idea of absolute rest. They suggest rather that . . . the same laws of electrodynamics and optics will be valid for all frames of reference for which the equations of mechanics hold good. We will raise this conjecture . . . to the status of a postulate, and also introduce another postulate, which is *only apparently* irreconcilable with the former, namely, that light is always propagated in empty space with a definite velocity c which is independent of the state of motion of the emitting body.

With these words Einstein proclaimed the universality of c and the equivalence of inertial frames for all purposes. It was a major development in the history of ideas. Hermann Bondi, the British cosmologist, has commented on it in a very interesting way[2]:

> If the spirit of the principles of Newtonian relativity is applied, it follows that this basic feature, the velocity of light, should be independent of the inertial system from which it is observed. The assertion that this is so . . . is known as the special theory

[1]A. Einstein, *Ann. Physik*, **17,** 891 (1905); translated by W. Perrett and G. B. Jeffery, 1923, in *The Principle of Relativity*, Dover, New York.

[2]H. Bondi, *Endeavour*, **20,** 121 (1961).

> of relativity . . . Thus the theory is, in a sense, profoundly conservative; the principles that have been found to hold for dynamics are stated to hold for the whole of physics. The special theory of relativity is a necessary consequence of any assertion that the unity of physics is essential, for it would be intolerable for all inertial systems to be equivalent from a dynamical point of view yet distinguishable by optical measurements. It now seems almost incredible that the possibility of such a discrimination was taken for granted in the nineteenth century, but at the time it was not easy to see what was more important—the universal validity of the Newtonian principle of relativity or the absolute nature of time . . . .

This last phrase of Bondi's brings us to the heart of Einstein's thinking. The nature of time was the crux. It took Einstein's insight to bring about the realization that a fundamental analysis of this kind was called for. But equally impressive was Einstein's conviction that all observable physical phenomena must depend only on *relative* motions. And before we follow Einstein into special relativity as such, it will be appropriate to say a little about relativity principles in general, and about the Newtonian scheme which, as Bondi points out, has so much in common with the newer dynamics of Einstein.

RELATIVITY ACCORDING TO GALILEO AND NEWTON

What is a relativity principle? It is an assertion about the laws of nature as they would be determined by observations made in different frames of reference. The assertion is made in an explicit form, founded upon experience, that there are whole classes of reference frames with respect to which the laws of physics have precisely the same form. Observers stationed in these different frames make measurements which may in themselves be different, but which lead to the same conclusions about laws of force, etc. A relativity principle is a statement, in other words, about an essential lack of privilege in nature; it places a whole group of reference frames on a precisely equivalent status. Bondi put it this way at the beginning of a review article on relativity[1]:

> It is the purpose of every physical theory to describe in a concise manner a wide variety of phenomena. In many cases this will necessitate, as part of the theory, a prescription for applying the theory to systems in differing states of motion. A prescription

[1]H. Bondi, *Rept. Progr. Phys.*, **22,** 97–120 (1959).

of this kind, a code of translation as it were, will generally consist of a mathematical system of transformation laws.

It is in the nature of transformation laws to change most quantities but to leave some quantities unchanged. These [latter] are called the invariants of the transformation and serve to define its character. A physical statement of what these invariants are is called a principle of relativity, and the fundamental equations of a theory usually define the principle of relativity applicable to it.

Relativity in this sense is already embodied in Newtonian mechanics, which suggests a unique importance and interest to be attached to inertial reference frames—i.e., to all those frames (related one to another by some constant velocity) in which Galileo's law of inertia holds good.

It is to Galileo that we owe the first clear discussion of such questions. In the famous book in which he advocated the Copernican view of the solar system against the Ptolemaic, he argued that the vertical path of a falling object does not compel one to the conclusion that the earth is stationary.[1] He gave, by way of analogy, the example of a rock dropped from the top of the mast of a ship. Whether the ship is at rest or moving with a constant velocity, the rock always lands just at the foot of the mast. Thus an observation of the point of impact on the deck reveals nothing of the ship's state of motion. The measured path of a falling rock, in the frame in which it is released from rest, is always a vertical straight line, whether that frame is stationary or has some constant horizontal velocity with respect to the earth. Expressed in this way, the result is an embodiment of a relativity principle, for it contains the assertion that a certain law of nature (the law of free fall) is the same in all reference frames differing only by some constant horizontal velocity.

A closely related example is the observation of one and the same falling object from two different reference frames. Suppose, for example, that a rock, released from a moving ship, is observed both from the ship and from the land. Then in one case the path would be seen as a vertical straight line, and in the other case as a parabola. But if, now, we analyze these motions with the help of Newton's dynamics, we discover that, although the motions are different, they yield identical conclusions about

[1]Galileo, *Dialogue Concerning the Two Chief World Systems—Ptolemaic and Copernican*, translated by Stillman Drake, Univ. California Press, Berkeley, 1953.

the magnitude of the acceleration and hence about the force causing it.

The relationship between measurements of a given motion as observed in different reference frames is expressed by a set of transformation equations. In Newtonian mechanics, the relevant equations are named in honor of Galileo, who first recognized the principle of inertia. They relate measurements of position, time, velocity, and acceleration in an inertial frame S to the corresponding measurements made in a frame S' that has a constant velocity v along the x direction relative to S:

$$\left.\begin{aligned} x' &= x - vt \quad (\text{and } x = x' + vt) \\ y' &= y \\ z' &= z \\ t' &= t \\ u_x' &= u_x - v \quad (\text{and } u_x = u_x' + v) \\ a_x' &= a_x \end{aligned}\right\} \begin{array}{l}\text{Galilean} \\ \text{transformation} \\ \text{equations}\end{array} \qquad (3\text{–}4)$$

Actually any follower of Newton would have regarded the first three equations alone as being sufficient to define the relationship between two inertial frames in relative motion. It would not have been considered necessary to spell out the relationship $t' = t$, since the notion of a universal time was in the very foundation of Newtonian mechanics: "Absolute, true and mathematical time, of itself, and from its own nature, flows equably without relation to anything external."[1] Given this universal time, and the first three of equations (3–4), the relation of the measures of velocity in different inertial frames is completely defined, and the measure of an acceleration is the *same* in all such frames. *In Newtonian mechanics, the acceleration is an invariant.*

We shall go a little further to see how the statement of a physical law of some kind is affected (or not) by the act of transforming it to a different reference frame. We shall then be ready to consider Einstein's ideas more fully.

THE TRANSFORMATION OF NEWTON'S LAW

The basic statement of Newtonian mechanics is $F = ma$. What do we mean by a transformation of this law? Unless we have an explicit law of force, $F = ma$ is only a definition of F. So

[1]Newton's *Principia*, Book I, Scholium after Definitions.

let us consider the force provided by the interaction between two bodies. Suppose it is a function only of their separation, and does not depend on their velocities or accelerations. Then we can put

$$F_{12} = f(x_2 - x_1) \tag{3-5}$$

This is the force exerted on body 2 by body 1. Thus the equation of motion of body 2, as observed in an inertial frame S, is given by

$$f(x_2 - x_1) = m_2 a_2 \tag{3-6}$$

With the help of the Galilean transformation, equations (3–4), we can rewrite Eq. (3–6) entirely in terms of measurements made in the frame S'. The transformed value of the force is defined by

$$F_{12}' = f(x_2' - x_1')$$

But we have

$$x_2' - x_1' = (x_2 - vt) - (x_1 - vt) = x_2 - x_1$$

Thus

$$F_{12}' = f(x_2 - x_1) = F_{12}$$

Turning now to the right side of Eq. (3–6), the Galilean transformation gives us $a = a'$, and in Newtonian dynamics the inertial mass m is a constant: $m_2 = m_2'$. Thus we are able to write

$$F_{12}' = f(x_2' - x_1') = m_2' a_2' \tag{3-7}$$

We see, then, in this admittedly very simple example, how the Newtonian law of motion is unaltered by the Galilean transformation. A more complicated force law, as long as it involved only the *relative* positions and velocities of two interacting particles, would possess this same property that we call invariance. If, however, the force depended on *absolute* positions and velocities—e.g., if it contained terms like $x_2^2 - x_1^2$—the form of the equation of motion would cease to be the same in all inertial frames. Nothing in our experience has revealed such a situation, which would make the laws of physics appear different in our laboratory and in a train or plane moving at constant velocity. Consider, for example, the simple phenomenon of a collision of two billiard balls as perceived by several observers in different inertial frames. Each observer analyzes this (same) collision process from his own perspective and records such data as position, velocity, etc., as measured with respect to his own frame. The different sets of data will, in general, not agree with each other, but the dynamic laws of the collision, as deduced from the measurements, are the same for all observers. If the physical laws were different for one or more observers, this might be a clue to the uniqueness of certain frames of reference, and thus all inertial frames would not be on an equal footing.

It was the assumption of 19th-century optics that a unique reference frame, defined by the luminiferous ether, did indeed exist, but we have seen something of the difficulties and contradictions that this assumption led to. We find no positive evidence for the existence of any such frame; it appears that all our physical experience is consistent with the dynamical equivalence of all inertial frames. It remains true, however, as we saw in Chapter 1, that the predictions of Newtonian mechanics are not borne out by experience if sufficiently high speeds are involved. What has gone wrong—or, to be more positive, what scheme of things will allow us to fit all this experience together? It turns out that a solution can be found in the following terms:

All inertial frames are indeed equivalent under all conditions, but the laws of motion and transformation must be modified.

What we shall now proceed to do is (following Einstein) to replace the Galilean transformations by another set—the Lorentz-Einstein transformations—that will allow us to preserve the Newtonian principle of relativity and yet encompass the deviations from Newtonian dynamical behavior. We shall begin with a restatement of Einstein's basic ideas.

EINSTEIN AND THE UNIVERSALITY OF c

The novel idea that Einstein brought into physics was very profound, yet perfectly easy to grasp. He pointed out that the analysis of motions had been based on an abstraction—the existence of a universal, absolute time. He argued that this was untenable, and that, as a matter of fundamental principle, "the justification for a physical concept lies exclusively in its clear and unambiguous relation to facts that can be experienced."[1] One may have metaphysical notions about time, but when it

[1]Quoted from a lecture by Einstein in 1921 (reprinted in his *Ideas and Opinions*).

comes to making measurements one is dealing with observations made with actual physical devices—we call them clocks. Here are some examples:

A mechanical clock or watch, with spring, escapement, and gears
The rotating earth
A beam of ammonia molecules in a maser
A quartz crystal
A group of radioactive particles
Your own pulse

If we wish to describe quantitatively the motion of a body, we must be able to give its coordinates as a function of time. What does this involve? Again we may quote from Einstein's 1905 paper, which states the case better than any paraphrase would:

> We have to take into account that all our judgments in which time plays a part are always judgments of simultaneous events. If, for instance, I say "That train arrives here at 7 o'clock," I mean something like this: "The pointing of the small hand of my watch to 7 and the arrival of the train are simultaneous events."

This seems straightforward—almost trivial. But then Einstein goes on to point out that a problem arises if, as in describing the motion of a body, we want to establish a time relationship between events that occur at different places. Suppose, for example, that we wish to measure the velocity of a body. We note its position $\mathbf{r}_1$ at time t_1 and its position $\mathbf{r}_2$ at time t_2. Then

$$\mathbf{v} = \frac{\mathbf{r}_2 - \mathbf{r}_1}{t_2 - t_1}$$

But, in terms of Einstein's analysis, this means that we must make use of a clock reading t_1 that is simultaneous with the arrival of the body at $\mathbf{r}_1$, and *another* clock reading t_2 that is simultaneous with the arrival of the body at $\mathbf{r}_2$. And whatever particular method of measurement we use, our observations have no meaning unless we can also define what we mean by the *same* time at the two different locations. If we could transmit information at infinite speed there would be no problem. But this we cannot do. Instead, we seize on the large but finite speed of electromagnetic signals in vacuum as the means for relating time measurements at different places. This was Einstein's argument, and it proceeded roughly as follows.

Suppose we have observation stations at two different places, A and B. A clock at A can record the time differences between events that occur in the immediate vicinity of A. Similarly, a clock at B can record the times of events that occur in the immediate vicinity of B. We thus have an "A time" and a "B time." To establish a *common* time for A and B together, we say, *as a matter of definition*, that *the time required for a light signal to travel from A to B is equal to the time required for a light signal to travel from B to A*. Thus if a light signal starts out from A at $t = 0$, is reflected by a mirror at B, and arrives back at A at $t = t_0$, then the time at which the signal reaches B is *defined* as being $t_0/2$. This gives us a specific procedure for synchronizing clocks that are at different places.

A corollary of this definition of the relation of time measurements at different places is that, again as a matter of definition, the speed of light in vacuum has the same value in all circumstances. Einstein incorporated this statement in his original formulation of the special relativity theory. His theory is entirely founded, in fact, on the following two postulates:

Postulate 1: All inertial frames are equivalent with respect to all the laws of physics.

Postulate 2: The speed of light in empty space always has the same value c.

It is very remarkable that a whole new dynamics can be based on two such brief statements. Only an extraordinary mind could have discerned this possibility.

THE SECOND POSTULATE AND OBSERVATIONAL EVIDENCE

Einstein's second postulate, on the universality of c, was a dramatic innovation. It was framed in a way that would seem to deny the possibility of any independent experimental check. It has been argued, however, that this essential feature of special relativity can in fact be based on observational evidence (much of it, however, not available until long after Einstein had developed the theory).[1] We shall comment briefly on some aspects of this point of view, in terms of specific experiments.

[1]This was discussed by H. P. Robertson in an article in a special issue of *Reviews of Modern Physics*, celebrating Einstein's 70th birthday [*Rev. Mod. Phys.*, **21**, 378 (1949)].

The Michelson-Morley experiment

In this experiment the Michelson interferometer was made with arms of equal length. No fringe shift occurred as the apparatus was rotated through 360°—an operation taking only a few minutes, during which time the state of motion of the laboratory did not change significantly. This null result is consistent with the proposition that the speed of light is the same in all directions with respect to a reference frame having some arbitrary (but unknown) motion through space.

The Kennedy-Thorndike experiment[1]

In 1932 R. J. Kennedy and E. M. Thorndike performed an experiment that was very similar to the Michelson-Morley experiment but differed from it in two significant ways. First, the arms of the interferometer were made of different lengths. (The difference was about 16 cm, as large as was consistent with getting good-quality interference fringes.) Second, the apparatus was fixed in the laboratory and the interference fringes were observed over a period of months—an experiment putting extreme demands on the mechanical stability and constancy of the apparatus. No fringe shifts associated with either the diurnal or the seasonal changes in the motion of the laboratory were found. If we interpret the Michelson-Morley experiment as we have done just above, then the null result of the Kennedy-Thorndike experiment can be construed as evidence that the time for light to travel the extra distance $2|l_2 - l_1|$ in the longer arm of the interferometer is the same in reference frames having quite different velocities. (It may be noted, strictly parenthetically, that if the contraction hypothesis were adopted as an explanation of the Michelson-Morley result, it would not explain the Kennedy-Thorndike experiment. If you refer to Eq. (3–2), which incorporates a Lorentz-Fitzgerald contraction, the time difference as calculated from the ether-wind hypothesis will still change if the speed v changes, except in the special case that $l_{10} = l_{20}$. This carries its own implication that our measures of time as well as of distance will need to be modified.)

Light from a moving source

It is an essential feature of Einstein's second postulate that the speed with which light signals travel away from a given source is

[1]R. J. Kennedy and E. M. Thorndike, *Phys. Rev.*, **42**, 400 (1932).

quite independent of any motion of the source itself. In our first discussion of the Michelson-Morley experiment (in Chapter 2) we stated that this was indeed the case. For a long time it was believed that this was proved by observations on the light from close binary stars. The two members of any such binary system have large relative velocities, and when one star has a component of velocity toward the earth the other will be moving away. It was argued that if these velocities were communicated to the emitted light, the *apparent* motions of the stars would be distorted away from the Newtonian orbits required by the law of gravitation. No such distortions were observed. It has been more recently argued, however, that since these binary star systems are usually surrounded by a gas cloud, which absorbs and then reradiates the light from the stars, the speed of the light that crosses interstellar space may in any case be independent of any possible influence of the original moving sources.[1] Subsequently, however, experiments have been made on rapidly moving terrestrial sources of radiation which verify this aspect of Einstein's second postulate in a convincing way. In one such experiment made with high-energy photons, not visible light, the source consisted of unstable particles (neutral π mesons) traveling at 99.975% of the speed of light. The measured speed of the photons emitted forward with respect to this motion was $(2.9977 \pm 0.0004) \times 10^8$ m/sec.[2] Reference to Table 1–2 will show that this is in excellent agreement with the best values of c obtained for stationary sources. In Chapters 5 and 6 we shall discuss in more detail the radiation from moving sources, in connection with the relativistic law of addition of velocities and related phenomena.

THE RELATIVITY OF SIMULTANEITY

An immediate consequence of Einstein's prescription for synchronizing clocks at different locations is that simultaneity is relative, not absolute. Let us see how this follows.

Suppose that three observation stations A, B, and C are equally spaced along the x axis of an inertial frame S in which they are all at rest. We can construct a simple x-t coordinate system, on which we draw "world lines" (to use the accepted phraseology) showing the development of the system in space

[1]J. G. Fox, *Am. J. Phys.*, **30,** 297 (1962).

[2]T. Alväger, F. J. M. Farley, J. Kjellman and I. Wallin, *Phys. Letters*, **12,** 260 (1964).

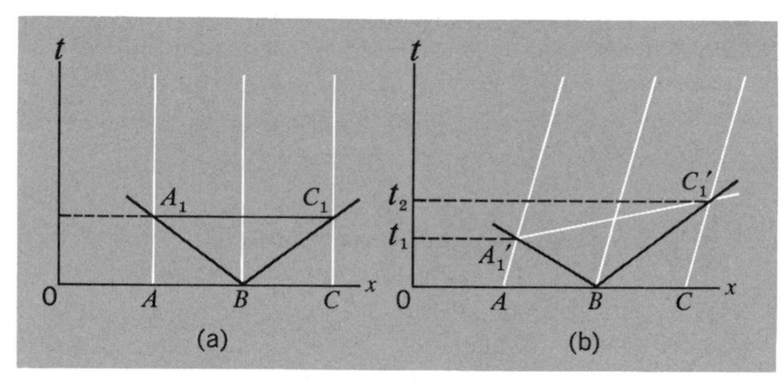

Fig. 3–2 (a) Space-time diagram showing experiment to define simultaneity at stations A and C (at rest in this reference frame) by light signals emitted from a station B midway between them. (b) Equivalent experiment for the case in which A, B, and C all have a velocity with respect to the reference frame.

and time [Fig. 3–2(a)]. The world line of any given particle is just a graph of its position as a function of time; it provides a complete picture of the history of the particle as observed within a given frame of reference. The world lines of A, B, and C are of course just vertical lines parallel to the t axis, corresponding to x = constant. Suppose that a light or radio signal is sent out from B at $t = 0$. It travels at the same speed c forward and backward along the x axis—an assertion that embodies the universality of c. This signal is described by two sloping lines $x = x_B \pm ct$. The arrival of the signal at the positions of A and C is thus given by the intersections A_1, C_1, and simultaneity at the positions of A and C is *defined* by the line A_1C_1, parallel to the x axis, which joins a series of points possessing the same value of t.

But now suppose that A, B, and C are at rest in an inertial frame S' which is moving with respect to S at a speed v along the x direction [Fig. 3–2(b)]. The world lines of A, B, and C are now inclined as shown. A signal sent from B at $t = 0$ is again described (in S) by the lines $x = x_B \pm ct$, and the arrival of the signal at the positions of A and C is now given by the intersections A_1' and C_1'. These are clearly not simultaneous for S, because the line $A_1'C_1'$ is manifestly not parallel to the x axis. Or, to put it more concretely, the signal reaches A before it reaches C because, as observed in S, A is running to meet the signal pulse whereas C is running away from it. But we require

B to be midway between A and C in S' as well as in S. Accepting the universality of c and the equivalence of inertial frames, we therefore *demand* that A_1' and C_1' represent simultaneous events in S'. (An *event*, from the standpoint of relativity theory, is completely characterized by its space and time coordinates in a given frame of reference.) *Our judgment of simultaneity is a function of the particular frame of reference we use.*

It is natural to ask why we should base this definition of simultaneity on the velocity c in particular and not on any other possible signal velocity. The simplest answer to this is to point to the obvious uniqueness of c, not merely as the speed of light, but as the ultimate speed in all of dynamics. A more convincing reply (at least in the long run) is that this choice does indeed have the consequence that every known physical law has the same form in all inertial frames.

In the above discussion we have demonstrated in a qualitative way the relativity of simultaneity. Our next step must be to develop the quantitative aspects of time and space measurements according to special relativity.

THE LORENTZ-EINSTEIN TRANSFORMATIONS

Look now at Fig. 3–3. It depicts the operation of defining simultaneity at stations A and C which are moving at speed v with respect to an inertial frame S. We have already discussed such a diagram (cf. Fig. 3–2). But now we have added lines to represent the coordinate axes of the frame S' in which A and C

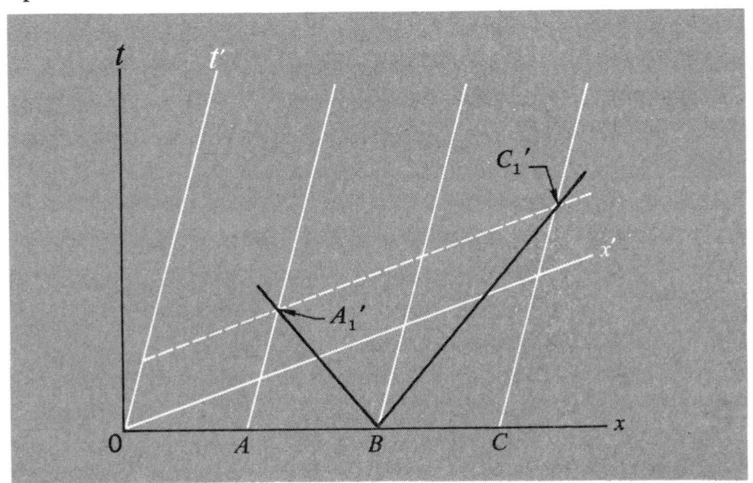

Fig. 3–3 Specification of coordinate axes (x, t) and (x', t') for two reference frames in relative motion.

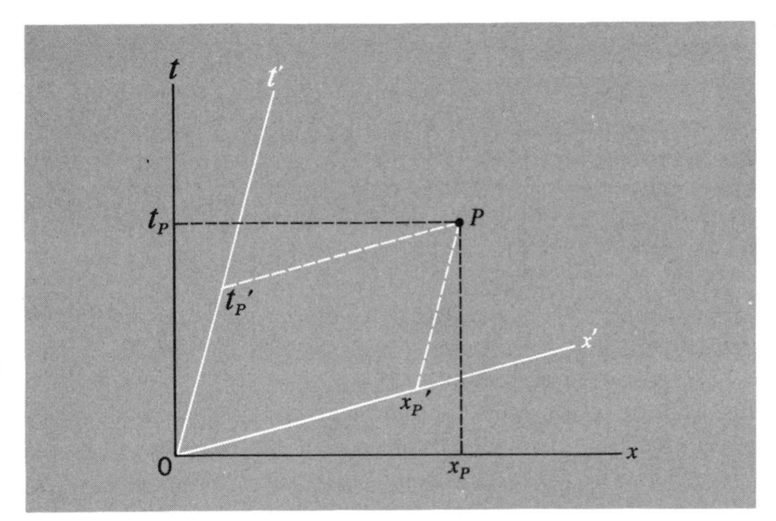

Fig. 3–4 Space-time coordinates of a given point event in two different inertial frames.

are at rest. How have we done this? The axis of t' is readily described; it is the line $x' = 0$, i.e., the world line of the origin of S'. And since the frame S' has a speed v along the x direction with respect to S, the position of this origin is described in S by the equation $x = vt$ if the origins of S and S' coincide at the time $t = 0$.

What about the axis of x'? This is the line that connects all points corresponding to $t' = 0$. Any line of the form $t' =$ constant is parallel to this x' axis. But the line $A_1'C_1'$ is just such a line, since A_1' and C_1' are events by which simultaneity in S' is defined. Hence we construct the x' axis by drawing a line parallel to $A_1'C_1'$, and for convenience we make it pass through O, which is thus described both by $x = 0$, $t = 0$ and by $x' = 0$, $t' = 0$. The noncoincidence of the axes of x and x' does not, of course, imply any geometric tilting of one with respect to the other; it is a purely formal tilting in the abstract space constructed from the x and t coordinates.

Now this type of diagram displays for us a key feature of the kinematic transformations of special relativity. In Fig. 3–4 any point P in the plane of the diagram represents what is called a *point event*, which can be characterized alternatively by the values of x and t or of x' and t'. And our construction implies that x' and t' alike should be linear functions of both x and t. Similarly, x and t are linear functions of x' and t'. This linearity is a fundamental property of the transformation equations. If they did not have this form, a motion recorded as motion at constant velocity along a straight line in one frame (say S) would not be recorded as uniform rectilinear motion in S'. This would therefore conflict with Galileo's law of inertia and with our basic dynamical condition that all inertial frames are equivalent.

The symmetry implied by the relativity principle means that the form of the relationships must be as follows:

$$x = ax' + bt'$$

with (3–8)

$$x' = ax - bt$$

These are set up so as to resemble as closely as possible the Galilean transformation [equations (3–4)] to which they must certainly reduce for sufficiently small values of the speed v of S' relative to S. The motion of the origin of S as measured in S' is defined by putting $x = 0$ in the first of these equations. Similarly, the motion of the origin of S' as measured in S is defined by putting $x' = 0$ in the second equation. The velocities are equal and opposite and both of magnitude v. This gives us the condition

$$b/a = v \qquad (3\text{–}9)$$

Next we consider the descriptions according to S and S' of a light signal traveling in the positive x direction. Let the signal originate at O of Fig. 3–3. It is then described by the following very simple equations in S and S', respectively:

$$x = ct \qquad x' = ct' \qquad (3\text{–}10)$$

Substitute these particular expressions for x and x' in equations (3–8), and we get the following:

$$\begin{aligned} ct &= (ac + b)t' \\ ct' &= (ac - b)t \end{aligned} \qquad (3\text{–}11)$$

Eliminating t and t' between these last equations, and using the condition $b = av$ from Eq. (3–9), we find

$$c^2 = a^2(c^2 - v^2)$$

Therefore,

$$a = \frac{1}{(1 - v^2/c^2)^{1/2}} \qquad (3\text{–}12)$$

It may be noted that this coefficient, a, is precisely the factor $\gamma(v)$ that emerged in our dynamical analysis in Chapter 1—cf. Eq. (1–22). We can now rewrite equations (3–8) in the following explicit form:

$$\left.\begin{aligned} x &= \frac{1}{(1 - v^2/c^2)^{1/2}}(x' + vt') = \gamma(x' + vt') \\ \text{and} \quad x' &= \frac{1}{(1 - v^2/c^2)^{1/2}}(x - vt) = \gamma(x - vt) \\ \text{where} & \\ \gamma(v) &= (1 - v^2/c^2)^{-1/2} \end{aligned}\right\} \tag{3–13}$$

These differ from the Galilean transformations by having the factor γ (≥ 1) as a multiplier on the right, and it is clear that the Galilean equations are a limiting form of equations (3–13) for $v/c \to 0$.

Given equations (3–13) it is a matter of elementary algebra to obtain the following expressions for t and t':

$$\begin{aligned} t &= \gamma(t' + vx'/c^2) \\ t' &= \gamma(t - vx/c^2) \end{aligned} \tag{3–14}$$

The reduction of these to the Galilean relation $t = t'$ requires $x \ll ct$ as well as $v/c \ll 1$. Equations (3–13) and (3–14) are the revised version of the transformations relating x and t for two inertial frames in relative motion along the x direction.

To complete our statement of the transformation of measurements between two reference frames in relative motion, we need the connection between measures of a distance (y or z) transverse to the direction of relative motion of the frames. Clearly, if space is isotropic, all displacements transverse to the unique direction defined by the relative motion are equivalent, and it is not difficult to conclude that the appropriate transformations are of simple equality:

$$y = y' \qquad z = z' \tag{3–15}$$

We can argue this on the grounds that, if it were *not* true, we should have a way of detecting absolute displacements and motions. Imagine, for example, that we constructed two identical square grids, representing xy coordinate systems, one of which is going to be associated with a frame S and the other with a frame S'. We assume that the grids are first checked for identity of spacing when at rest relative to each other. We suppose that they are then set in relative motion along x, and that paint brushes mounted every 10 cm along the y axis of S leave stripes on the grid S'. Likewise, paint brushes mounted every 10 cm along y' leave stripes on S. Each inertial frame thus receives its

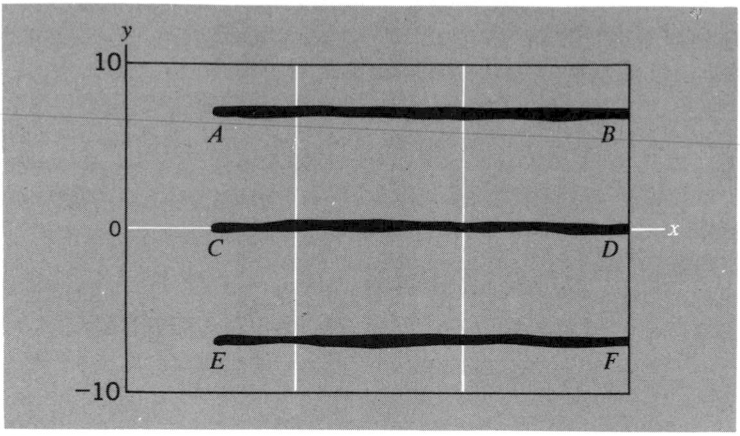

Fig. 3–5 Hypothetical experiment to compare transverse scales of distance (y) for two frames in relative motion along x.

own permanent record of the encounter. We cannot conceive that the stripes would be anything but straight lines parallel to x; otherwise we should have, in effect, a detector of absolute position along x. And suppose the grid S received stripes as shown in Fig. 3–5 from the brushes mounted at $y' = 10$ cm, 0 cm, and -10 cm. We should then be forced to conclude, not only that the 10-cm intervals of S' have shrunk as measured in S, but also that the 10-cm intervals of S appear larger than 10 cm in the record they leave on S'. But this would mean an asymmetry depending on the direction of motion—whether it was to the right or to the left along x. Such a result would violate our essential ideas of relativity and of the homogeneity and isotropy of space. We conclude, therefore, that the measure of a transverse distance (y or z) must be the same for all inertial systems that are in relative motion along x, as expressed in equations (3–15).

For convenience and future reference, we give the complete set of transformations below, expressed both ways—i.e., S' coordinates in terms of S, and vice versa:

THE LORENTZ-EINSTEIN TRANSFORMATIONS

$$
\begin{aligned}
x' &= \gamma(x - vt) & \qquad x &= \gamma(x' + vt') \\
y' &= y & \qquad y &= y' \\
z' &= z & \qquad z &= z' \\
t' &= \gamma(t - vx/c^2) & \qquad t &= \gamma(t' + vx'/c^2)
\end{aligned}
$$

with $\gamma = (1 - v^2/c^2)^{-1/2}$, where v is the velocity of S' as measured in S

(3–16)

The above array of transformation equations was introduced by H. A. Lorentz in 1904 as a basis for modifying electromagnetic theory so as to reconcile the null result of the Michelson-Morley experiment with the existence of a unique inertial frame provided by the luminiferous ether. But Einstein discovered the equations quite independently a year later with the help of his fresh and radical approach to the whole problem.

MORE ABOUT THE LORENTZ TRANSFORMATIONS[1]

In deriving the Lorentz transformations in the last section, we considered only the requirements imposed by light signals traveling along the x direction. A more general approach would have developed them by applying the requirements of Einstein's second postulate to a light signal traveling in an arbitrary direction. Having already set up the transformations, however, we can use them to illustrate a seeming paradox which is contained in Einstein's postulate. It is this: Suppose that a burst of light begins spreading out (in vacuum) from the origin of frame S at $t = 0$. At any later time t the light will have reached all points on a sphere of radius r, centered on the origin of S, such that $r = ct$. Then if this same phenomenon is observed with respect to a frame S', moving with respect to S with any velocity v, the description of the expanding burst of light is again a sphere, in this case centered on the origin of S'—even though, by definition, the origins of S and S' coincide only at the instant $t = t' = 0$.

To see how this result emerges, we take the equation $r = ct$ and rewrite it in terms of position and time coordinates measured in S'. By first squaring both sides of the equation we get

$$x^2 + y^2 + z^2 = c^2t^2$$

Now use the right-hand set of equations (3–16). The above equation then becomes the following:

$$\gamma^2(x' + vt')^2 + (y')^2 + (z')^2 = \gamma^2c^2(t' + vx'/c^2)^2$$

It may be noted that the cross terms in $x't'$ on the two sides of the equation are equal, and so disappear. Collecting the other terms, we have

$$\gamma^2(x')^2(1 - v^2/c^2) + (y')^2 + (z')^2 = \gamma^2(t')^2(c^2 - v^2)$$

[1]Having once recognized that these transformations were arrived at by both Lorentz and Einstein, we shall usually in future refer to them by this briefer and more customary title.

But

$$\gamma^2(1 - v^2/c^2) = 1$$

Therefore,

$$(x')^2 + (y')^2 + (z')^2 = c^2(t')^2$$

which defines a sphere of radius r' such that

$$r' = ct'$$

This result, which at first sight appears to do violence to one's commonsense ideas, is bound up with the relativity of simultaneity. Points which, as measured in S, are reached at the *same* time t, are reached at *different* times as measured in S', in such a fashion that the light is properly described as lying on a spherical shell expanding at speed c in both frames.

MINKOWSKI DIAGRAMS: SPACE-TIME

A valuable aid to the arguments in this chapter has been the use of graphs, with axes representing position and time, which allow one to display the complete history of a one-dimensional motion. The use of such graphs in special relativity was introduced by H. Minkowski in 1908, and they are customarily referred to as Minkowski diagrams. On any such diagram, as we have seen, any individual event—e.g., a light signal striking a detector, or one tick of a watch—is uniquely represented by some point P (Fig. 3–6). The detailed specification of this event, however, in terms of numerical values of x and t, can be made in infinitely many different ways according to the particular reference frame chosen. The description of a point event is described in frame S by the coordinates (x, t) and in S' by the coordinates (x', t'). If the origins of S and S' are chosen so as to coincide at $t = t' = 0$, then the relation between (x, t) and (x', t') is contained in the Lorentz transformations of equations (3–16).

It is very convenient to use ct, rather than t, to describe the time coordinate. Both coordinates, ct and x, are then expressed as distances, and if the scale of distance is chosen to be the same for both, the world line of a light signal starting out at $x = 0$, $t = 0$, is a bisector of the angle between the axes. This holds good in all reference frames. We can represent any one such frame (say, S) by drawing the axes of x and ct at right angles to one

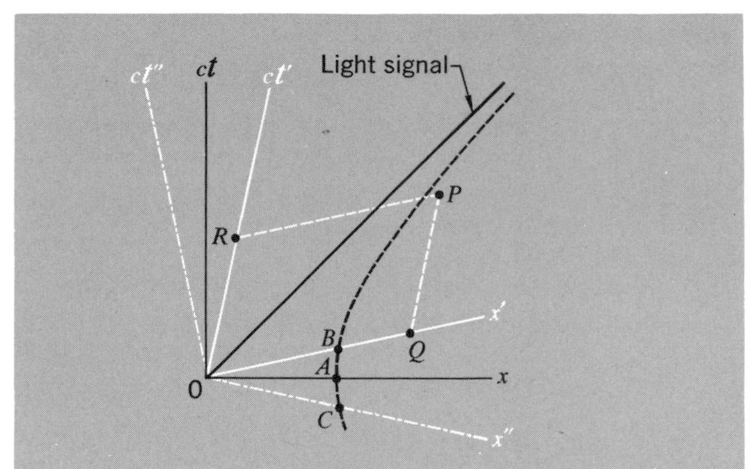

Fig. 3–6 Minkowski diagram, showing three different coordinate systems and a calibration hyperbola to define unit distance along x for each system.

another. Other reference frames (S' and S'', for example) are then characterized by nonorthogonal axes for distance and time. (In Fig. 3–6, S' has some positive velocity along x with respect to S, and S'' has a negative velocity.) There is nothing special or privileged about that frame which we choose to show with orthogonal space and time axes.

To read off the space and time coordinates of a given point event P, we draw lines through it, parallel to the space and time axes of any chosen reference frame, and read off the intercepts. In Fig. 3–6 we have taken S' as the frame, so OQ gives the measure of x' and OR the measure of ct'. It is very important to realize, however, that in a diagram such as Fig. 3–6, *the scale, representing unit distance, is not the same along the different axes* x, x', x'', *etc.* To be specific, if we draw the rectangular hyperbola defined by

$$x^2 - (ct)^2 = 1 \tag{3–17}$$

its intersections A, B, and C with the axes of x, x', and x'' will in each case define unit distance from O for the particular frame in question (see Fig. 3–6). The justification of this procedure is embodied in the discussion in the next section.

A SPACE-TIME INVARIANT

Let us now return to the basic kinematic relations of special relativity, as expressed in the Lorentz transformations. We have seen how special relativity was born out of Einstein's recognition that neither "where" nor "when" has any absolute meaning. Can we salvage something that *is* well defined out of this description of things? The answer is yes. Suppose that some identifiable event takes place at position x and time t as measured in a system S. Another observer, moving with respect to the first in a system S', will record it as having occurred at x' and t' such that

$$x' = \gamma(x - vt)$$
$$t' = \gamma(t - vx/c^2)$$

Let us evaluate $(ct')^2 - (x')^2$:

$$\begin{aligned}(ct')^2 - (x')^2 &= \gamma^2[(ct - vx/c)^2 - (x - vt)^2] \\ &= \gamma^2[(c^2 - v^2)t^2 - (1 - v^2/c^2)x^2]\end{aligned}$$

i.e.,

$$(ct')^2 - (x')^2 = (ct)^2 - (x)^2 = s^2 \tag{3–18}$$

where the quantity s^2 is the same for all inertial observers—i.e., an invariant.

The calibration of the axis scales in a Minkowski diagram follows at once from this result. For if we put $s^2 = -1$, we have from Eq. (3–18) the dual statements

$$\begin{aligned}x^2 - (ct)^2 &= 1 \\ (x')^2 - (ct')^2 &= 1\end{aligned} \tag{3–19}$$

If we choose to take x and ct as orthogonal axes, as in Fig. 3–6, the first of the above two equations defines the rectangular hyperbola of that figure. The x axis of S is the line $ct = 0$, and we see that the intersection of this line with the right branch of the hyperbola defines $x = 1$. But, in an exactly similar way, the line $ct' = 0$ defines the axis x' of S', and from the second equation of (3–19) we can deduce that the intersection of this line with the same hyperbola corresponds to $x' = 1$. This hyperbola thus acts as a calibration curve for all the inertial frames that may be represented on a single Minkowski diagram. What it amounts to is that the hyperbola of Eq. (3-17), which defines a particular relation between x and t for an infinite number of different events as described in the single frame S, is also the locus in space-time of all events representing time zero and x coordinate equal to unity in different inertial frames.

What significance does the invariant s^2 have? Dimensionally, its square root s is a distance, so it is tempting to regard s as some kind of a measure of the distance (in space-time) by which a given event is removed from the origin of space-time coordinates. One can see, however, that s^2 may be either positive or negative (or, of course, zero) so that s itself may be either real or imaginary. Any interpretation of the physical significance of s must, therefore, be a rather formal one. By introducing ict as a variable one can convert equations (3–19) into statements of essentially the same form as the Pythagorean theorem. This then paves the way for talking about time as the fourth dimension, and so on. But such developments really add very little to the basic statements of space-time transformations, and we shall not pursue them at this point. We shall note only that s^2, because it is an invariant, is a useful quantity in the characterization of events, and that $s^2 = 0$, in particular, defines the world line of a light signal that passes through the space-time origin ($x = 0, t = 0$).

PROBLEMS

3–1 Verify that a Lorentz contraction along the direction of motion leads to a time difference that is independent of orientation for the light beams in a Michelson interferometer with arms of unequal length [cf. Eq. (3–2)].

3–2 In the Kennedy-Thorndike experiment, as described in the text, the difference between the lengths of the two arms of the interferometer was about 16 cm.

(a) If the sun is assumed to be at rest in the ether, the velocity of the apparatus at any instant is the vector sum of the earth's orbital velocity $\mathbf{v}$ (30 km/sec) and the surface velocity $\mathbf{u}$ due to the earth's rotation (about 0.5 km/sec at the equator). Show that the maximum fringe shift δ ($= c\,\Delta t/\lambda$) that might be expected during the course of 1 day, according to Eq. (3–2), is given by

$$\delta \approx \frac{4uv}{c^2}\,\frac{l_{10} - l_{20}}{\lambda}$$

Evaluate the magnitude of δ numerically for $\lambda = 6 \times 10^{-7}$ m.

(b) Astronomical observations suggest that the sun has a velocity of about 220 km/sec relative to the center of our galaxy. If the galactic center is assumed to be at rest in the ether, what maximum fringe shift might the Kennedy-Thorndike experiment have revealed in the course of 1 year?

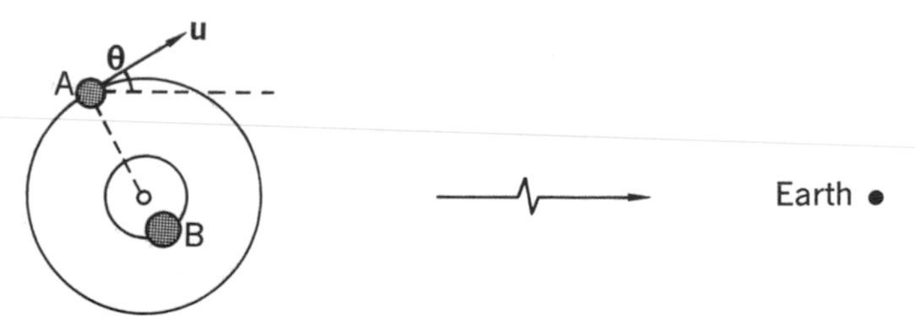

3–3 The figure shows a double-star system with two stars, A and B, in circular orbits of the same period T about their center of mass. The earth is in the plane defined by these orbits at a distance R of many light-years. Let the speed of A in its orbit be u; then at any instant it has a velocity v_r ($= u\cos\theta$) along the line from the double-star system to the earth. When light emitted from A reaches the earth, its observed Doppler shift (change of wavelength of characteristic spectral lines) reveals the value of v_r at the instant of emission.

(a) If the speed of light from A to the earth were modified by the motion of A, so as to be equal to $c + v_r$, show that the value of v_r, *as inferred from spectroscopic observations on earth*, would *appear* to be varying with time in accordance with the following equation if $u \ll c$:

$$v_r = u \sin\frac{2\pi}{T}\left(t - \frac{R}{c} + \frac{Rv_r}{c^2}\right)$$

(One might wonder how, if we are uncertain about the nature of light and its speed, we can justifiably infer the speed of a star from the Doppler shift of its spectral lines. The answer is that we have confidence in such measurements from our experience with the Doppler shift observed directly for terrestrial sources with known velocities.)

(b) For the double-star system Castor C, $u = 120$ km/sec, $T = 0.81$ day, and $R = 48$ light-years. Does the above theory predict that the observed radial velocity as a function of time should deviate appreciably from the equation $v_r = u\sin 2\pi(t - R/c)/T$? Experimentally, A. H. Jay and R. F. Sanford, *Astrophys. J.*, **64,** 250 (1926), found that it did not. (After J. H. Smith, *Introduction to Special Relativity*, Benjamin, New York, 1965.)

3–4 Given that

$$x' = \gamma(x - vt)$$

and

$$t' = \gamma(t - vx/c^2)$$

derive the equations for x and t in terms of x' and t'.

3–5 An event occurs at $x' = 60$ m, $t' = 8 \times 10^{-8}$ sec in a frame

S' ($y' = 0, z' = 0$). The frame S' has a velocity $3c/5$ along the x direction with respect to a frame S. The origins of S and S' coincide at $t = 0, t' = 0$. What are the space-time coordinates of the event in S?

3–6 The space and time coordinates of two events as measured in a frame S are as follows:
Event 1: $x_1 = x_0$, $t_1 = x_0/c$ ($y_1 = 0$, $z_1 = 0$)
Event 2: $x_2 = 2x_0$, $t_2 = x_0/2c$ ($y_2 = 0$, $z_2 = 0$)

(a) There exists a frame in which these events occur at the same time. Find the velocity of this frame with respect to S.

(b) What is the value of t at which both events occur in the new frame?

3–7 Frame S' has a speed $v = 0.6c$ relative to S. Clocks are adjusted so that $t = t' = 0$ at $x = x' = 0$.

(a) An event occurs in S at $t = 2 \times 10^{-7}$ sec at a point for which $x = 50$ m. At what time does the event occur in S'?

(b) If a second event occurs at (10 m, 3×10^{-7} sec) in S, what is the time interval between the events as measured in S'?

3–8 Two inertial coordinate systems S and S' move with speed $c/2$ with respect to each other. Draw a Minkowski diagram relating these two systems. (Let the axes of x and ct for S be at right angles in your drawing.)

(a) Draw calibration hyperbolas that allow you to define unit distances along the axes of x, x', ct, and ct'.

(b) Plot the following point events on the diagram: (1) $x = 1$, $ct = 1$; (2) $x' = 1$, $ct' = 1$; (3) $x' = 2$, $ct' = 0$; (4) $x = 0$, $ct = 2$.

(c) From your diagram determine the coordinates in S' (or S) corresponding to the above events.

3–9 (a) Convince yourself that the equations describing a rotation of coordinates (x, y) through an angle θ are

$$x' = x \cos\theta + y \sin\theta$$
$$y' = -x \sin\theta + y \cos\theta$$

(b) Show that the Lorentz transformation corresponds to a rotation of coordinates (x, ict) through an angle $\theta = \tan^{-1}(i\beta)$, where $\beta = v/c$.

A number of other problems involving the use of the Lorentz transformations will be found at the end of Chapter 4.

In classical physics it was always assumed that clocks in motion and at rest have the same rhythm, that rods in motion and at rest have the same length. If the velocity of light is the same in all coordinate systems, if the relativity theory is valid, then we must sacrifice this assumption. It is difficult to get rid of deep-rooted prejudices, but there is no other way.

A. EINSTEIN AND L. INFELD,
The Evolution of Physics (1938)

4

Relativity and the measurement of lengths and time intervals

IN THE previous chapters we have sought first to show the need for a revision of kinematics and dynamics, and then to present the essentials of Einstein's ideas. We shall now begin our study of the application of his theory to a number of dynamical questions, and a few prefatory remarks may be appropriate.

In the first encounter with Einstein's relativity, one may get impressions like these:

1. This is something quite apart from the mechanics of Galileo and Newton.

2. There is a lot of algebraical wizardry—much of it bewildering. One can learn to do some of the tricks, but it doesn't make much physical sense.

Such feelings are very natural. No matter how long one has lived with the results of special relativity, there is something very nonintuitive about it. None of our everyday, direct experience involves speeds greater than $10^{-6}c$ (riding in a jet aircraft). Even an earth satellite passenger only gets to $2 \times 10^{-5}c$. But relativity *does* make good sense, and is *not* in a separate compartment from Galileo-Newton. Keep these thoughts at the back of your mind as you study it. And wherever possible, check to see how the equations of motion and transformation according to special relativity reduce to Newtonian ones for velocities that are small compared with c.

OBSERVERS

The literature of relativity is full of references to *observers*, whose role is to make judgments on the positions and times at which events occur. (We introduced such hypothetical observers ourselves at one stage in the last chapter.) Almost always the observer is portrayed as being at rest with respect to one or other of two frames; by imagining an observer in each frame, one can picture an actual process for obtaining two different space-time descriptions of the same event. All this seems both harmless and reasonable. One may even speculate, with Bishop Berkeley, whether it means anything to speak of an event in the absence of someone to observe it![1] Nevertheless, the use of this language contains certain dangers. It is very important to realize, as Einstein in essence pointed out, that the role of an observer is simply to record coincidences, i.e., pairs of events which occur at the same space-time point. A clock reading at a particular point in a given frame of reference is an event in this sense; indeed, our concern in this chapter is overwhelmingly with events of this kind. The things that we more familiarly think of as physical events—e.g., the collision between two objects, or the emission of a photon by an atom—are, for our purposes, to be regarded as happenings that coincide with events describable as readings on clocks. The incessant references to clock readings may well seem artificial and somewhat wearisome, but they do serve to emphasize an absolutely essential feature—that we are dealing with a very explicit problem of *measurement*.

To prevent or dispel some possible misconceptions, we add the following specific comments.

1. Although an event is by definition represented by a single point in space-time, it may nevertheless leave an enduring record of itself. A criminal touches a glass, for example, and leaves a fingerprint. The touching of the glass is an event, occurring at a unique place and a unique time in a given frame of reference.

[1]Of course, physicists in general accept the reality of any event that is *in principle* observable.

A second later, even if the glass has not been moved, the fingerprint is at a different point in space-time. But it remains as a record that a certain event took place. A still more pertinent example would be of a watch that falls onto a concrete floor and stops dead. If it is left where it fell, it represents a permanent record of the "watch-strikes-floor" event. And an observer, coming upon the scene long afterward, can note down the space and time coordinates of the event as measured in a reference frame defined by the floor (for space coordinates) and by the hands of the watch (for time coordinate).

2. The last remark above should make it clear that an observer is not necessarily limited to making measurements in a reference frame to which he himself is attached. One can appeal to all kinds of familiar experience that embody this fact. For example, one is a passenger in a train that shoots through a station. On the platform is a sign with an arrow pointing in the train's direction of motion and carrying the words "New York 10 miles." Just above the sign is a station clock that reads 10:53 A.M. As an observer attached to a certain reference frame, defined by the train, one can nevertheless record the space-time coordinates of an event—"train passes through station"—as measured in the relatively moving reference frame of the station and the ground to which it is attached.[1] Very often, however, one will see statements such as the following: "An observer A in frame S observes that an event occurs at position x and time t; the same event is observed to occur at position x' and time t' by an observer B in frame S'." What is really being said here is just that the event has space-time coordinates (x, t) in one frame and (x', t') in the other. But there is conjured up a picture of an observer, cloistered in his own particular frame of reference, unable to record anything except the measures of position and time in that frame. Our example of the passenger in a train shows how unnecessarily restrictive this is. The passenger can note not only the reading on the station clock but also the reading on his own watch. Of course there *may* be situations in which an observer is limited to making observations on instruments in his own inertial frame, but it is still not precluded that he should receive information about the results of observations made in some other frame. Such communications are indeed a commonplace.

[1]We are treating as negligible the time taken for light to pass from station clock to passenger.

3. The last and most treacherous aspect of introducing an observer attached to a given frame of reference is that one may get the impression that this observer has some kind of bird's-eye view of the whole of his reference frame at a given instant. *This is entirely false.* A single observer is not ubiquitous; at a given instant he has awareness only of events occurring at his own location—e.g., a burst of photons striking his retina. If those photons constitute the visual image of a distant object, then clearly they represent the object as it was at an earlier time—earlier by the transit time of the light itself. An observer at a given point may well be able to collect information from which he can construct a description of what *was* happening at a given earlier instant in his frame of reference, but he *cannot* have this complete picture at the instant itself. One must be immediately on guard if one reads such colloquialisms as: "An observer attached to frame S sees the event as happening at position x and time t," or "To an observer in frame S it looks as if. . . ." Almost always, these statements are simply statements about the space-time coordinates of a particular event as established by measurements in frame S. If one really means to talk about looking or seeing, then an extra feature—the transmission of information from one point to another—is involved.

The purpose of this discussion, then, is to focus the attention where it belongs—on the specification of point events according to the measures of space and time in given frames of reference. If we mention observers, it will, unless otherwise stated, merely be in terms of their role as recorders of these measurements.

POINT EVENTS AND THEIR TRANSFORMATIONS

The concept of an individual point event, as introduced in Chapter 3, is basic to the successful use of relativistic kinematics. Any given event, as we have remarked, is essentially transitory—it is represented by a single point in space-time. But the space-time coordinates of the event are different as measured in different inertial frames. This is, of course, as true in the Galilean-Newtonian scheme as in special relativity. But the Einstein-Lorentz transformations have superseded the Galilean ones, and reduce to them, as we have seen, if the value of c can be taken as effectively infinite.

Everything that we do in this chapter is based on the Lorentz transformations. For convenience, therefore, we shall restate them in the usual form:

$$\left.\begin{aligned} x' &= \gamma(x - vt) & x &= \gamma(x' + vt') \\ y' &= y & y &= y' \\ z' &= z & z &= z' \\ t' &= \gamma(t - vx/c^2) & t &= \gamma(t' + vx'/c^2) \end{aligned}\right\} \tag{4–1}$$

with

$$\gamma = (1 - v^2/c^2)^{-1/2}.$$

If an event occurs at the coordinates (x, y, z, t) in S, and at (x', y', z', t') in S', the measurements in S and S' are related as above. The left-hand set of equations conveniently describes the transformation from S to S', and the right-hand set describes the corresponding transformation from S' to S. Given either set of transformations, the other set follows. The equations (4–1) are purely and simply the recipe for relating two descriptions of the same point event.

If we have two point events, each of them can be represented by its space-time coordinates in S and/or S'. For one space dimension we can draw an x-t graph and mark in the individual point events as in Fig. 4–1. And we have:

Event 1:

$$\left.\begin{aligned} x_1' &= \gamma(x_1 - vt_1) & x_1 &= \gamma(x_1' + vt_1') \\ t_1' &= \gamma(t_1 - vx_1/c^2) & t_1 &= \gamma(t_1' + vx_1'/c^2) \end{aligned}\right\}$$

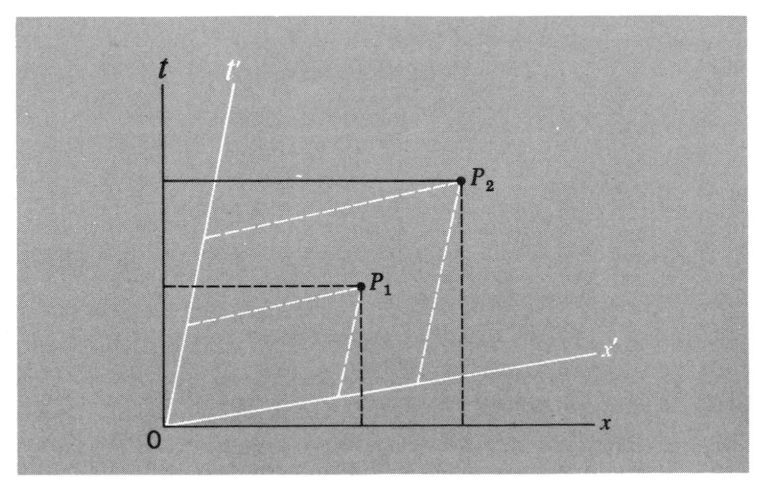

Fig. 4–1 Two different point events, each describable in either S or S′.

Event 2:

$$\left.\begin{aligned} x_2' &= \gamma(x_2 - vt_2) \qquad & x_2 &= \gamma(x_2' + vt_2') \\ t_2' &= \gamma(t_2 - vx_2/c^2) \qquad & t_2 &= \gamma(t_2' + vx_2'/c^2) \end{aligned}\right\}$$

We can then evaluate the separation of the events in space and time in either frame. Thus

$$x_2' - x_1' = \gamma[(x_2 - x_1) - v(t_2 - t_1)] \tag{4–2}$$

$$t_2' - t_1' = \gamma[(t_2 - t_1) - v(x_2 - x_1)/c^2] \tag{4–3}$$

Example. Frame S' has a speed $v = 0.6c$ relative to S. Clocks are adjusted so that $t = t' = 0$ at $x = x' = 0$. Two events occur. Event 1 occurs at $x_1 = 10$ m, $t_1 = 2 \times 10^{-7}$ sec ($y_1 = 0, z_1 = 0$). Event 2 occurs at $x_2 = 50$ m, $t_2 = 3 \times 10^{-7}$ sec ($y_2 = 0, z_2 = 0$). What is the distance between the events as measured in S'?

First, we have

$$v^2/c^2 = \tfrac{9}{25}$$

and hence

$$\gamma = (1 - v^2/c^2)^{-1/2} = \tfrac{5}{4}$$

Then

$$\begin{aligned} x_2' - x_1' &= \tfrac{5}{4}[(50 - 10) - \tfrac{3}{5}(3 \times 10^8)(3 - 2)10^{-7}] \\ &= 27.5 \text{ m} \end{aligned}$$

TIME MEASUREMENTS

Here is where our intuitions chiefly play us false. There is no such thing as *the* time. "The time" is not a metaphysical abstraction, it is the reading on a clock. If all measurements were made in a single inertial frame, there would be no problem. We could synchronize clocks everywhere by a radio-signal technique, and that would be that. But applying this technique to two inertial frames in relative motion shows us that the synchronization of clocks at different places is not absolute; each frame has its own criterion of simultaneity. It is from this that we get the transformation equations for time measurements:

$$t' = \gamma(t - vx/c^2) \qquad t = \gamma(t' + vx'/c^2)$$

These equations in effect say the following:

Imagine we have two sets of identical clocks—infinitely many in each set—placed along a straight line (the axis of x or x'). All the clocks in S are set to read the same time t at what is judged to be the same instant in S. All the clocks in S' are similarly synchronized according to the criterion of simultaneity in S'. But the S' clocks all have a velocity v (along x) relative to the S clocks—and the S clocks all have the velocity $-v$ relative to the S' clocks. Then:

1. Measurements made (in S) of the readings of the S' clocks at a specified instant t (in S) will show a progressive change of reading (t') with position (x) as measured in S.

But similarly:

2. Measurements made (in S') of the readings of the S clocks at a specified instant t' (in S') will show a progressive change of reading (t) with position (x') as measured in S'.

Note the emphasis on measurement—at a particular point, at a particular time, in a particular frame.

Example. What is the time difference, as measured in S', between the same two events considered in the example in the previous section?

Using Eq. (4–3), we have

$$t_2' - t_1' = \gamma\left[\left((t_2 - t_1) - \frac{v}{c}\frac{x_2 - x_1}{c}\right)\right]$$

$$= \frac{5}{4}\left[(3 - 2)10^{-7} - \frac{3}{5}\left(\frac{50 - 10}{3 \times 10^8}\right)\right]$$

$$= 2.5 \times 10^{-8}\ \text{sec}$$

In the above examples we have taken two arbitrary and unrelated point events and shown how to calculate their separation in position or time in another reference frame. But now we shall take up a more specific and more subtle type of question. How does one identify, in some arbitrary frame of reference, a pair of point events whose separation in space may justifiably be taken as a measure of the length of a particular object, or alternatively whose separation in time may justifiably be taken as a measure of the time interval between successive ticks of a particular clock? It is here that we shall see, in vivid form, the consequences and implications of the revised description of events in space-time. We are brought face to face with the question of what it really

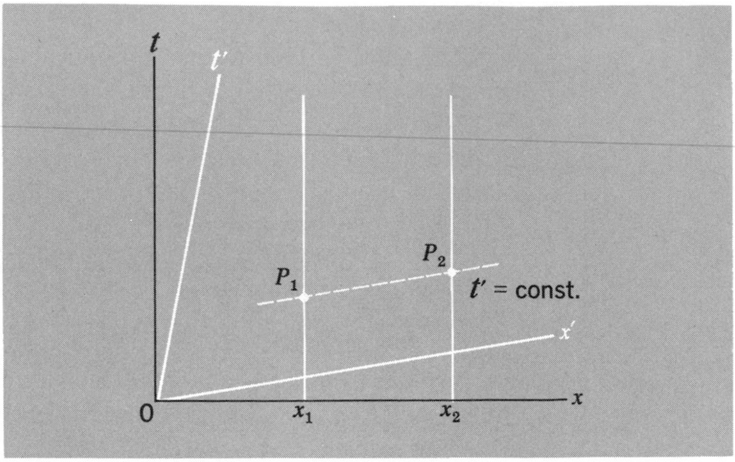

Fig. 4–2 *Lorentz contraction. The spatial interval between the ends of an object at rest in S is measured in S′ at a time t′.*

means to make a measurement, once the tacit and unsupportable assumptions of classical mechanics are stripped away. And from this analysis there will emerge the celebrated results that a moving object appears to be contracted and that a moving clock appears to run slow.

THE LORENTZ CONTRACTION

An object at rest in an inertial frame S (x, y, and z all constant) is continually changing its time coordinate. It is described by an infinite succession of space-time events. A point particle at rest in an inertial frame S is described in space-time by a line made up of all the point events that have the same values of (x, y, z) in S.

Thus in Fig. 4–2 the two lines parallel to the t axis, and representing $x = x_1$ and $x = x_2$, could be the world lines of the two ends of a body whose length in S is always measured as l_0:

$$l_0 = x_2 - x_1$$

Suppose now that we want to define a procedure for finding the length of the body by means of measurements made in some other frame S'. What should be done? The answer is very straightforward: Measure the positions of the two ends of the body at the same time t' as judged in S'. Now any line representing $t' =$ constant (Fig. 4–2) will intersect the world lines of the ends of the body at two points in the x-t diagram. These two point events (P_1 and P_2) have coordinates (x_1', t') and (x_2', t') with the same t' in each case. The length l of the object as measured in S' is given by

$$l = x_2' - x_1'$$

Now as measured in S these same two point events have x coordinates equal to x_1 and x_2, respectively, independent of the time as measured in S. Thus, using the Lorentz transformations, we have

$$x_1 = \gamma(x_1' + vt')$$
$$x_2 = \gamma(x_2' + vt')$$

Therefore,

$$x_2 - x_1 = \gamma(x_2' - x_1')$$

Hence

$$l = l_0/\gamma = l_0(1 - v^2/c^2)^{1/2} \qquad (4\text{–}4)$$

which corresponds exactly to the Fitzgerald-Lorentz contraction that was proposed to account for the null result of the Michelson-Morley experiment. You must understand precisely what l represents. It is the distance between the ends of an object at the same instant as judged in the particular frame S'. The criterion of simultaneity is different in any other frame, and hence the result of measuring the length will be different, too. This may seem strange at first, but according to Einstein's ideas it is inherent in the only kinds of observations that are physically meaningful.

(*Note:* Do not be misled by a casual inspection of Fig. 4–2, which might suggest that l, as represented by the line P_1P_2, is longer than l_0, as represented by $x_2 - x_1$. Remember the discussion of Minkowski diagrams in Chapter 3, in which it is pointed out that unit distance is represented by lines of different length along x and x' in any such diagram. When this is taken into account, the measure of l is always *less* than the measure of l_0.)

TIME DILATION

As we have already remarked, it is primarily in connection with time that our intuitions let us down. The most dramatic manifestation of this is in the phenomenon known as *time dilation.*

The time-dilation experiment

A sequence of pictures from the film "Time Dilation—an Experiment with Mu-Mesons," Education Development Center, Newton, Massachusetts. The text is adapted with permission from the film script.

We're on top of a mountain, Mt. Washington in New Hampshire. We shall count the number of mesons that arrive here and then we'll go down later to sea level and count the number that survived to arrive there. By comparing these numbers, we will show that these moving clocks run slow. To start, we need to detect these mu mesons.

(Prof. D. H. Frisch)

Every time a mu meson passes through the plastic scintillator, a flash of light is emitted. We detect them with the photomultiplier, which turns the flashes into electrical signals. If a mu meson stops and decays in the scintillator it makes a second flash, and we have arranged the oscilloscope so it shows only these decay pulses. The light from the decay pulses on the oscilloscope screen is picked up by the second photomultiplier. The register counts the mu mesons that have entered the scintillator and decayed. We're also interested in where the decay pulses occur on our time scale, and the easiest way to record that is with the Polaroid camera. [*Diagram after Frisch and Smith, "Measurement of the Relativistic Time-Dilation Using Mu-Mesons,"* Am. J. Phys., **31**, *342 (1963).*]

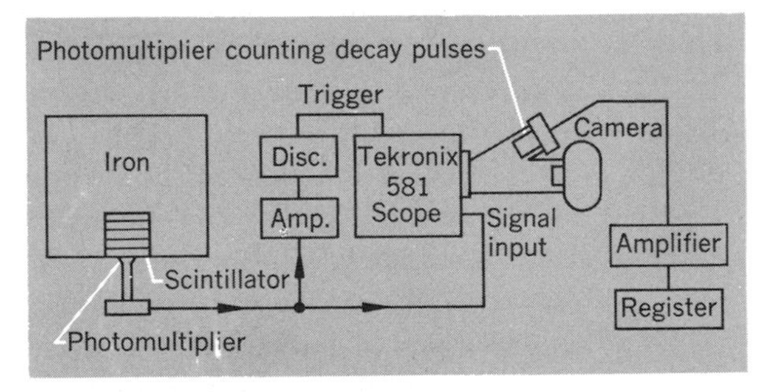

We are measuring the decay times of many mesons and plotting them on the chart. This particular meson lived for 4.85 μsec. and on the chart we have drawn a line representing the length of the decay time of that meson.

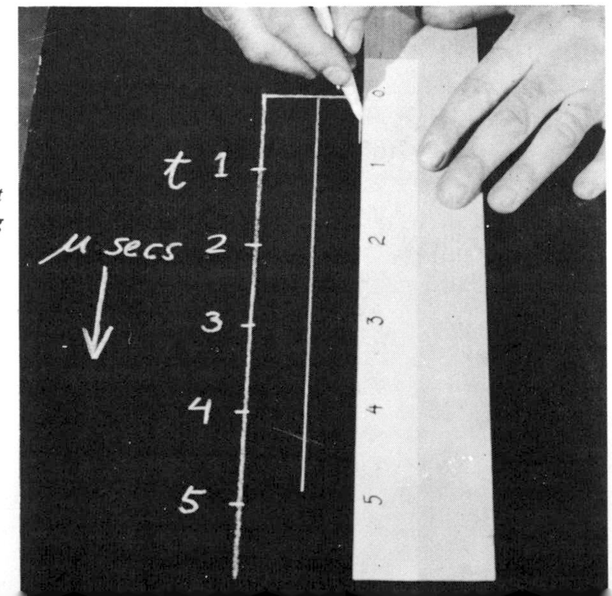

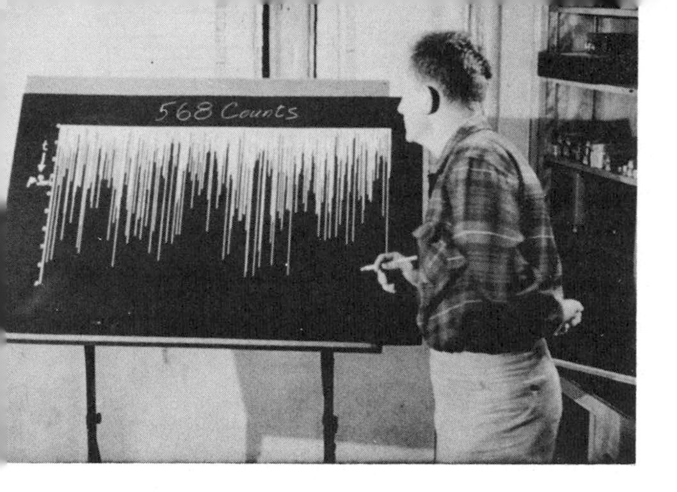

To see the pattern of the distribution of decay times we shall need to take many hundreds of counts, so we have made several 1-hour runs. Here's a finished chart based on 568 counts that we got during 1 hour.

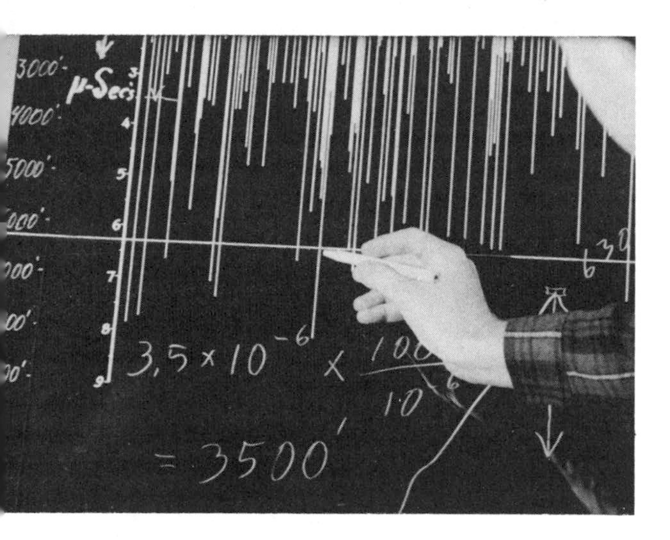

Since we know the mesons travel at nearly the speed of light, we can relabel the axis in thousands of feet instead of in microseconds and then ask how many mesons would live to reach sea level, 6300 ft down. Let me put a string across here at 6.3 μsec or 6300 ft and count how many mesons would reach this level. There are 27, based on the assumption that mu mesons decay when they're in flight in the same way as they do when they are at rest with respect to us. On this basis, if we take our equipment down to sea level we expect to find 27 mesons stopping and dying in the plastic scintillator each hour.

We counted 568 mesons on top of Mt. Washington, and now we've come down to sea level to see how many are left. We're at Cambridge, Massachusetts, counting the survivors of the mesons incident at the 6000-ft level. Let's take a full hour's count. Four-o-nine, four ten, four eleven, . . . , four twelve . . .— that's the hour. Instead of 27 we have 412 mesons left at sea level. 412 corresponds to only about 0.7 μsec on our meson-decay-distribution clock. 0.7 divided by 6.3 equals $\frac{1}{9}$. These mesons moving by us at 0.99 the speed of light keep time at $\frac{1}{9}$ the rate they do when they're at rest with respect to us.

The essence of the phenomenon is that a lapse of time as recorded on a *single* clock is compared to the results of measurements in a reference frame relative to which this clock is moving. In any such comparison, it emerges that the elapsed time as measured on the single clock is less than the difference between the two clock readings that describe the beginning and end of this interval in the other frame. The scale of time as measured by the single clock is apparently stretched out; hence the phrase time dilation as a description of it.

To see how this remarkable feature is inherent in special relativity, imagine that a single clock is at rest at the point $x = x_0$ in frame S. Consider two events corresponding to two different readings of the clock:

$$\text{Event 1: } (x_0, t_1) \qquad \text{Event 2: } (x_0, t_2)$$

Let us now calculate the time coordinates of these events as measured in the frame S' that has a velocity v with respect to S. Using the Lorentz transformations we have

$$t_1' = \gamma(t_1 - vx_0/c^2)$$
$$t_2' = \gamma(t_2 - vx_0/c^2)$$

Therefore

$$t_2' - t_1' = \gamma(t_2 - t_1)$$

If the difference $t_2 - t_1$ is written τ_0 and $t_2' - t_1'$ is written τ, then we have

$$\tau = \gamma\tau_0 = \frac{\tau_0}{(1 - v^2/c^2)^{1/2}} \qquad (4\text{–}5)$$

Compare Eq. (4–5) with (4–4). We see that the measured *length* of a body is *greater* in its rest frame than in any other frame, whereas the *time difference* between the events represented by two readings of a given clock is *less* in the rest frame of the clock than in any other frame.

The time-dilation phenomenon loses most of its mystery once we recognize that it is basically the consequence of comparing successive readings on a given clock with readings on *two different clocks*. Thus the result expressed by Eq. (4–5) involves in an essential way not only the rate of a given clock but also the procedure for *synchronizing* different clocks. The measurements that we have described require three clocks, two of which are in the same inertial frame but at different space coordinates in that frame. We can imagine all three clocks to be exactly alike—e.g., quartz oscillators governed by exactly the same dynamic equations in all inertial frames. If sitting at rest in a given frame they would continue to operate at identical rates. Observation of the time dilation requires that two of the clocks, relatively at rest but separated, be set to record $t = 0$ at the same instant as measured in their frame. This then involves Einstein's definition of simultaneity via the exchange of light or radio signals. Our use of the Lorentz transformations assumes that these synchronizations have been carried out, and the time dilation as an observational phenomenon follows.

OBSERVATION OF TIME DILATION WITH COSMIC-RAY MESONS

The very remarkable result represented by Eq. (4–5) cuts the ground from under the Galilean transformations. Expressed in colloquial terms, it says, "Moving clocks run slow." But even after all the discussion in the last section, one can scarcely help asking the question: Is such a bizarre result really true? The answer is yes, and we can point to unequivocal evidence for the phenomenon in the decay of rapidly moving particles.

Radioactive decay is one of those many phenomena in which individually random processes build up a well-defined picture. For any given type of unstable particles, it is possible to measure what fraction of a sample of them, first observed at a time t_0, still survives at a later time t. One can then predict that, within the statistical uncertainties of the finite numbers involved, the same result will be obtained with any other sample of the same type of particles. In this specific, statistical sense, unstable particles are a kind of clock; the measured survival fraction provides information about the elapsed time. This fact has, of course, been widely used in the study of our historic and geologic past through measurements on various long-lived isotopes that occur in nature. But it is an implication of Eq. (4–5) that if we take two groups of unstable particles of the same type, giving one group a speed v and letting the other group remain at rest with respect to our own frame of reference, the radioactive clock represented by the moving group will run slow compared to that of the stationary group. In a given time as measured by us, there will be fewer decays in the moving group than we should otherwise expect, and if v is close to c, the effect may be a large one.

A classic experiment on this time-dilation phenomenon was performed by B. Rossi and D. B. Hall in 1941,[1] using the μ mesons (muons) produced by cosmic rays entering the earth's atmosphere from outer space. More recently a filmed version of the same investigation has been made.[2] The essential facts relating to the experiment are as follows:

1. A μ meson is a charged particle that decays into an electron (+ or −), a neutrino (ν), and an antineutrino ($\bar{\nu}$):

$$\mu^{\pm} \rightarrow e^{\pm} + \nu_1 + \bar{\nu}_2$$

(It should perhaps be mentioned in passing that ν_1 and $\bar{\nu}_2$ are of different character, though these neutrinos will not concern us here.)

2. As produced by cosmic rays, the mesons travel predominantly downward through the earth's atmosphere with speeds very close to c.

3. With the help of charged-particle detectors one can record the arrival of a μ meson and, at a measured time later, the production of the energetic electron resulting from the decay process. Observation of the second stage means that the meson has been stopped in the detector, so that the decay of mesons at rest is being recorded.

4. From stage 3 one can build up a statistical record of the time intervals between arrival and decay for a large number of mesons.

5. Accepting the result of stage 4, one can predict what fraction of a group of mesons should be lost through decay in a trip of a given distance (l) and duration ($\approx l/c$) through the atmosphere (after making allowance for the removal of mesons by collision with atoms of the atmosphere).

6. Measurements are made of the rate of arrival of muons at the top of a mountain and at sea level. The mesons survive the downward journey in far greater numbers than one would predict from stage 5.

Let us look at some of the data. Table 4–1 shows a measured time distribution of the decay of a group of μ mesons at rest. In the filmed experiment, the counting apparatus recorded the arrival of 563 muons/hour (on the average) at an altitude of

[1]B. Rossi and D. B. Hall, *Phys. Rev.*, **59**, 223 (1941).

[2]Film, *Time Dilation—An Experiment with μ-Mesons* by D. H. Frisch and J. H. Smith, Education Development Center, Newton, Mass., 1963. See also D. H. Frisch and J. H. Smith, *Am. J. Phys.*, **31**, 342–355 (1963).

TABLE 4–1: MUON DECAY AT REST

Elapsed time, μsec	*No. of muons surviving*
0	568
1	373
2	229
3	145
4	99
5	62
6	36
7	17
8	6

about 2000 m. At a speed $\approx c$, mesons take about 6.5 μsec to reach sea level from this altitude, so that according to Table 4–1 the meson counting rate would have fallen to about 25 per hour at sea level. When the measurement was made, however, the result was in excess of 400 counts/hour. According to the clock represented by the moving mesons themselves, the journey lasted less than 1 μsec—actually only about 0.7 μsec! A very abbreviated pictorial version of the filmed experiment is shown on pages 98 and 99.

The time-dilation factor of about 9 in the above observations corresponds, of course, to a particular value of v. Using Eq. (4–5) we have

$$1 - v^2/c^2 = (\tau_0/\tau)^2 \approx \tfrac{1}{81}$$

which gives

$$v/c \approx 0.994$$

It should perhaps be pointed out that the muons produced by cosmic rays have a wide range of energies; the selection of a particular value of v/c is achieved by making observations only on those mesons that stop within a relatively thin layer of material (a plastic scintillator) after having traversed a specified large thickness of matter (iron + atmosphere). Muons of less than the chosen speed are stopped before they reach the plastic; those of higher speed pass right on through, and in both cases they go unrecorded, because to be accepted a muon must come to rest in the plastic scintillator and decay there.

As the above experiment shows, the time dilation is not only a fact of experience but can become a very large effect for clocks that are moving at speeds close to c with respect to a given reference frame.

ANOTHER INTERPRETATION OF THE TIME-DILATION EXPERIMENT

In discussing the time-dilation phenomenon as exhibited by μ mesons, we have considered everything from the standpoint of a frame of reference attached to the laboratory. In this frame the clock represented by a group of fast-moving mesons has been shown to run slow. But what if we try to describe the situation according to measurements made in an inertial frame that moves along with the mesons? In this frame the mesons are decaying at rest. As seen from this other viewpoint (e.g., from one of the mesons) the earth and its atmosphere are rushing upward at a speed almost equal to c. First a mountaintop flashes past, and then, a little later, the ground arrives. Now if frames in uniform relative motion are equivalent, the decay data for muons at rest in the laboratory (Table 4–1) must be applicable to muons at rest in our new frame. But we cannot change the result of the experiment, which is that about 75% (rather than 5%) of the mesons survive the journey from the mountaintop to sea level. How is this apparent inconsistency resolved? The answer is that, from the point of view of the moving mesons, the distance between mountaintop and sea level is strongly contracted. Let us make this quantitative.

Suppose that the vertical distance traveled by the mesons is H as measured by ordinary surveying—i.e., in the frame S attached to the earth. The mesons have a speed v down through the atmosphere. The duration Δt of the journey as measured in S is H/v ($\approx H/c$). According to the time-dilation equation (4–5) the journey takes a time $\Delta t' = \Delta t(1 - v^2/c^2)^{1/2}$ as measured in the meson rest frame S'.[1] But this time must also be measurable as a certain vertical distance H' (measured in S') divided by the speed with which the atmosphere moves upward past the mesons. This speed is v, because it is an essential feature of any relativity theory that there should be agreement about the magnitude of the relative velocity of two frames; only the sign of the velocity

[1]Note that, in deriving Eq. (4–5), the single clock (as represented now by a group of mesons) was assumed to be at rest in frame S, rather than in S', but the statement of the essential result is always the same—that the single clock records a smaller elapsed time than do the clocks in the other frame.

changes according to one's point of view. Thus we have

$$\Delta t = \frac{H}{v}$$

$$\Delta t' = \frac{H'}{v}$$

Therefore,

$$\frac{H'}{H} = \frac{\Delta t'}{\Delta t} = (1 - v^2/c^2)^{1/2}$$

$$\text{or} \quad H' = H(1 - v^2/c^2)^{1/2} \tag{4-6}$$

In other words, the distance between the top of the mountain and sea level is modified by the Lorentz contraction [Eq. (4-4)] as judged by measurements made in the reference frame of the mesons. We see then that a result upon which observers in all frames agree—what fraction of a group of unstable particles survives between one point event and another—may be attributed to time dilation or to Lorentz contraction, according to one's point of view.

The preceding discussion should help to reemphasize the fact that the time dilation is an expression of the definition of simultaneity, on the one hand, and of a particular type of measurement, on the other. To describe it by simply saying "Moving clocks run slow" may be convenient, but is also somewhat glib and can be misleading. For one thing, this statement suggests, quite contrary to relativistic ideas, that there is something absolute about motion. And, equally unfortunately, it suggests that some essential change occurs in the operation of the clock itself, that the physical basis of its operation has somehow been modified, whereas it is a central feature of relativity theory that just the opposite is true—that the operation of the clock as described in its own frame of reference is completely unaffected. We must recognize that whenever we speak of an object as moving, that statement has meaning only with respect to some given frame of reference (usually our own). As long as this is borne in mind, it is legitimate to speak of moving clocks or moving meter sticks. But beware!

MORE ABOUT TIME AND LENGTH MEASUREMENTS

In reading these last few sections, you have probably become aware that a special interest and importance attaches to a frame of reference that is fixed (at least in our imagination) to a particular object and always moves with it. This is the *rest frame*, and measurements made in it are called *proper*. No retreat from relativistic ideas is involved here. Although for each observer or object there is a uniquely defined rest frame, there is no extrinsic reason for preferring one over another. The length of a body as measured in its rest frame is called its proper length; we shall denote this by l_0. Any other measurement of the length is called nonproper, and, as we have seen, yields a value less than l_0 by a factor corresponding to the Lorentz contraction. Using the same terminology, we introduce the notion of *proper time*. This is time as measured always at some fixed point in a particular frame of reference. The measurement of a time interval between two events is nonproper unless both events are recorded in terms of the same clock. Referring back again to the muon-decay experiment, we can describe the time-dilation phenomenon as a manifestation of the distinction between proper and nonproper time intervals. And it will perhaps be instructive to rederive the time dilation and length contraction results by applying Einstein's second postulate (on the universality of c) to some hypothetical but very specific observations involving proper and nonproper measurements.

We shall consider a clock that is at rest in a frame S' and is moving at constant velocity with respect to another frame S. What sort of clock should it be? Well, it doesn't matter. If different types of clock responded in different ways to uniform motion, we should have, right there, a means of detecting such motion without reference to the outside world. This would in effect be a determination of absolute velocity, and would be at variance with the principle of relativity. So we have a free choice, and will take a clock whose action is particularly easy to analyze. It consists of a box containing two mirrors between which a light pulse bounces back and forth, and a dial that records one count at each return of the pulse.[1] If the distance between mirrors (i.e., the proper length) is l_0, the interval between successive counts is $2l_0/c$ of proper time. Suppose that such a clock moves transversely to its length at speed v with respect to some other inertial frame S (Fig. 4–3). The path of the light pulse with respect to this frame is ABC, and takes a time Δt as mea-

[1]This idealized clock has existed in the literature of relativity for a number of years. It bears more than a passing resemblance to one arm of a Michelson interferometer.

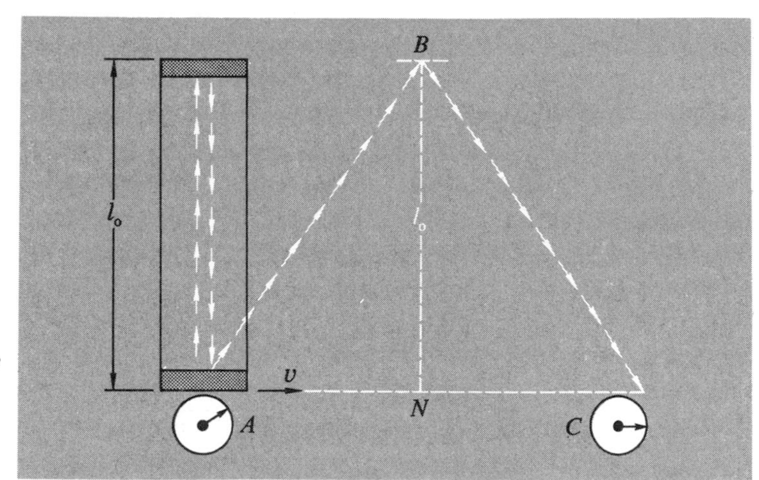

Fig. 4–3 Light-pulse clock, used to illustrate time dilation.

sured by the difference of readings of previously synchronized clocks (at A and C) at rest in S. We can argue the invariance of transverse lengths just as we did in Chapter 3; thus we have

$$AN = NC = v\,\Delta t/2$$
$$BN = l_0$$

Therefore,

$$AB + BC = 2\,[l_0^{\,2} + (v\,\Delta t/2)^2]^{1/2}$$

But the speed of light is the same for all observers, which means that the total distance traveled by the light pulse (as observed in the frame S) must be equal to $c\,\Delta t$. Hence

$$c\,\Delta t = 2\,[l_0^{\,2} + (v\,\Delta t/2)^2]^{1/2}$$
$$\Delta t = \frac{2l_0}{(c^2 - v^2)^{1/2}}$$

But the proper time interval $\Delta t'$ as measured on the moving clock is just $2l_0/c$. Therefore

$$\Delta t = \frac{\Delta t'}{(1 - v^2/c^2)^{1/2}} \tag{4–5a}$$

which represents the time-dilation formula once again.

In a very similar way we can argue the length contraction result. Suppose now we imagine our light-pulse clock to be traveling in the direction of its length at speed v with respect to some other frame S. Again the proper time for one round trip

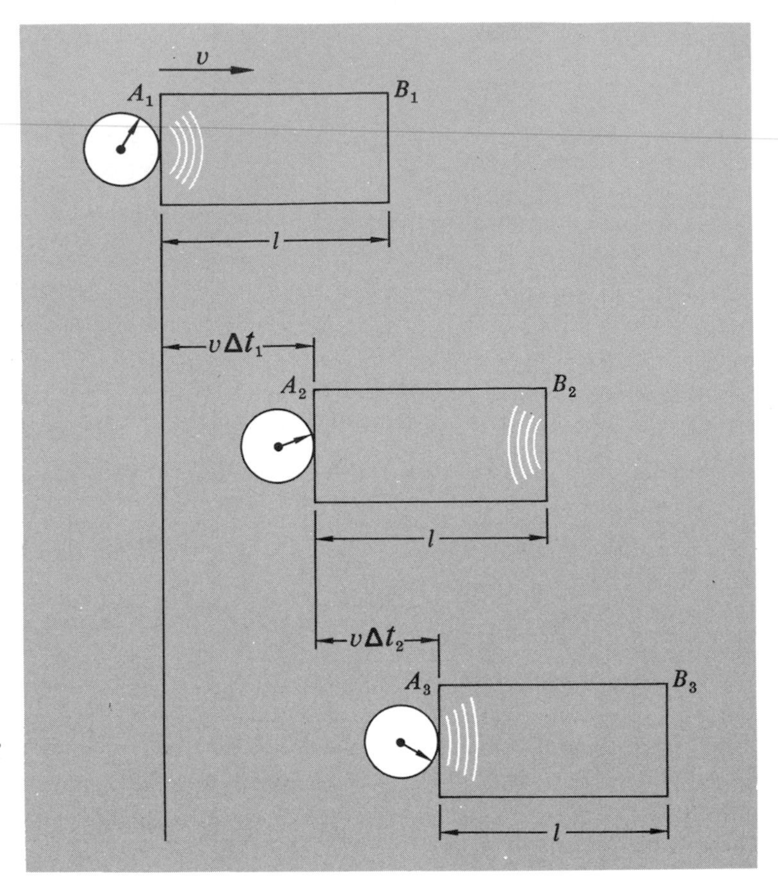

Fig. 4–4 Light-pulse clock, used to illustrate the Lorentz contraction.

of the light is $2l_0/c$, and we can use Eq. (4–5a) to tell us the elapsed time as measured in S. But we can calculate this time from the length of the light path as measured in S. To do this, we suppose that the length of the clock is l as measured in S (Fig. 4–4). Let the light take a time Δt_1 to travel from A_1 to B_2 (at which point it is reflected) and a time Δt_2 to travel back from B_2 to A_3. Then

$$l + v\,\Delta t_1 = c\,\Delta t_1$$

$$l - v\,\Delta t_2 = c\,\Delta t_2$$

$$\Delta t = \Delta t_1 + \Delta t_2 = \frac{2lc}{c^2 - v^2} = \frac{2l/c}{(1 - v^2/c^2)}$$

But, using Eq. (4–5a), we also have

$$\Delta t = \frac{2l_0/c}{(1 - v^2/c^2)^{1/2}}$$

Combining the last two equations we thus get

$$l = l_0(1 - v^2/c^2)^{1/2}$$

which reproduces Eq. (4–4).

The above analysis is, of course, no more than a description, in terms of a concrete example, of the Lorentz transformations as they apply to time and length measurements.

Let us make one last comment about the problem of recording point events that may occur anywhere within a given frame of reference. Having set up a space-coordinate grid, we must provide a means of measuring the time at which any event occurs. The most satisfactory way of doing this is to have time recorders (clocks) thickly scattered over the whole grid (say at every intersection of the grid lines); otherwise corrections must be made for the time taken for information to be relayed from a given point on the grid to some central recording point. We require, of course, that all the time recorders have been adjusted to read $t = 0$ simultaneously. But why should one always have to invoke the Einstein procedure of exchanging light or radio signals and allowing for the transit time? Why not simply check all the recorders against each other at some central point and then carry them to their individual stations? Time dilation is the answer. The recorders must themselves be treated as moving clocks during their passage from the origin to other locations; hence they will be judged to have lost time when compared at their destinations with clocks synchronized by the radio-signal technique. The error is $\Delta t - \Delta t'$ and is thus, by Eq. (4–5a), almost equal to $\frac{1}{2}v^2 \Delta t/c^2$ for $v/c \ll 1$. Once brought to rest at its new location, a transported clock will proceed to run at the normal rate, but its initial error persists. Since $\Delta t = s/v$, where s is the distance from the origin to any particular point, the error for any particular value of s is proportional to v and can be made arbitrarily small if v/c is made small enough. In other words, this method of standardizing time over a frame of reference is indeed defensible, but only under properly defined conditions.

A MICHELSON-MORLEY EXPERIMENT WITH LASERS

It is implicit in the discussion of light-pulse clocks in the preceding section that if we had two similar clocks of this type, oriented at right angles to one another, they would keep perfect time relative

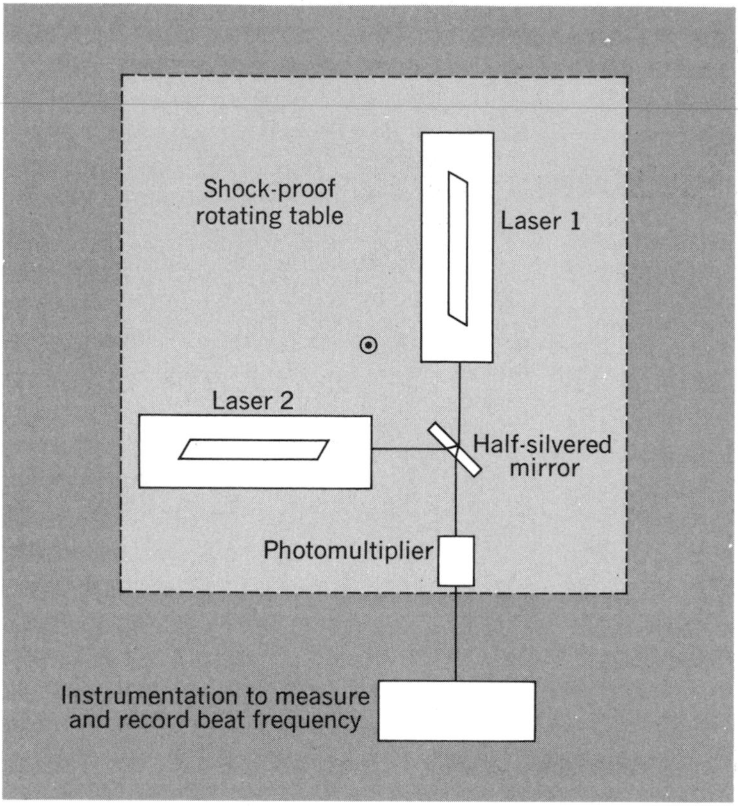

Fig. 4–5 Experimental arrangement to compare the frequencies of two lasers oriented at right angles to one another—a modernized version of the Michelson-Morley experiment, as performed by T. S. Jaseja et al.

to one another, and both would appear to go slow to the same degree from the standpoint of a frame with respect to which they had some velocity. In effect, this is a description of a Michelson interferometer and the null result of the Michelson-Morley experiment. But an even more complete analog has been constructed with two lasers, in a modern test of the basis of special relativity.[1] The essentials of the apparatus are shown in Fig. 4–5.

Although the frequency at which a laser operates must lie within the width of an atomic spectral line (in this case in the neon atom) the precise frequency is controlled by the speed of light and the distance between the mirrors at the two ends of the laser.[2] Thus the laser is in a very real sense a light-pulse clock. In the experiment, the characteristic frequency of both lasers was about 3×10^{14} cps (with a band width of only about 20 cps!).

[1]T. S. Jaseja, A. Javan, J. Murray, and C. H. Townes, *Phys. Rev.*, **133,** A1221 (1964).

[2]More specifically, the frequency is equal to $nc/2L$, where L is the length of the laser and n is a large integer. See if you can explain how this condition arises.

The difference of frequencies (the beat frequency) between the two lasers was recorded as the apparatus was rotated through 90°. No change of beat frequency (beyond what could be ascribed to local magnetic fields, etc.) was detectable within the accuracy of the measurement (about ± 3 kc/sec). This was less than 1/1000 of the change that one would calculate from an ether-wind hypothesis (cf. Chapter 2) and represents the most sensitive test yet made of the isotropy of space with respect to the speed of light signals.

RELATIVITY IS TRULY RELATIVE

Despite the fact that the relativistic transformations are founded on the complete equivalence of inertial frames, there may still seem to be something paradoxical about the length-contraction and time-dilation phenomena. Suppose we have two identical meter sticks in relative motion along the direction of their length. If I am riding in the rest frame of one of them I shall measure the length of the other to be less than 1 m. How, then, can I avoid the conclusion that my meter stick appears more than 1 m long to an observer in the rest frame of the other stick? You need never be perplexed by such a question if you once realize that each inertial frame must speak for itself, as it were. Observations made in one frame provide a description of events from the standpoint of that frame alone. The observations that lead A to say that B's meter stick is shrunk are A's observations; the conclusions drawn from those observations are valid for A but not for B. B can make his own observations—similar to A's, but not the same—and his conclusion is that A's meter stick is contracted. Both are correct. There is no contradiction, because it is physically impossible for both sets of observations to refer to one and the same reference frame.

This reciprocity can be nicely illustrated with the help of Minkowski diagrams. In Fig. 4–6 we show (twice) the representation of two reference frames S and S' related in the usual way. We suppose that there are two objects, each of proper length unity, one of which is at rest in S and the other at rest in S'. In Fig. 4–6(a) we show the world lines of the ends of the first object. Placing the origin at the left end, these world lines are the ct axis and the line AQ. Similarly, in Fig. 4–6(b), we show the world lines of the ends of the second object, the ct' axis and

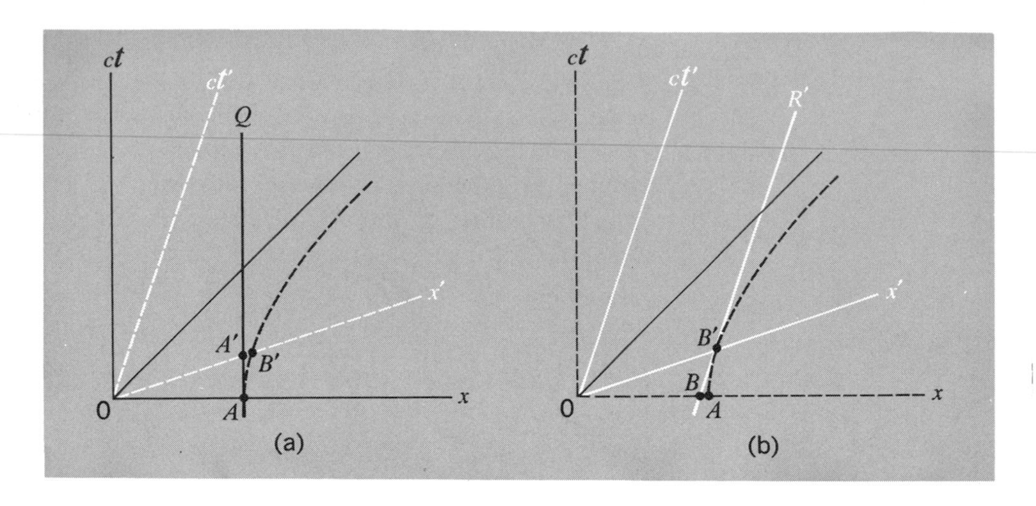

Fig. 4–6 Use of Minkowski diagrams to demonstrate the symmetrical nature of the Lorentz contraction. (a) A body of unit length (OA) at rest in S appears to be of less than unit length in S′. (b) A body of unit length (OB′) in S′ appears to be of less than unit length in S.

the line $B'R'$. Both diagrams show the calibration curve defined by the equation

$$x^2 - (ct)^2 = (x')^2 - (ct')^2 = 1$$

This curve, as we discussed in Chapter 3, intersects the x axis of any given reference frame at what corresponds (in that frame) to unit distance from the origin. Thus, as shown in both parts of Fig. 4–6, the curve passes through the point A on the x axis, such that $OA = 1$, and through the point B' on the x' axis such that $OB' = 1$.

Now in Fig. 4–6(a) the world line AQ is tangent to the calibration hyperbola at A, and in Fig. 4–6(b) the world line $B'R'$ is tangent to the calibration hyperbola at B'. The reason for this is that, in these Minkowski diagrams, the time axis for any frame is parallel to the tangent drawn at the point where the calibration hyperbola cuts the corresponding x axis. To see this, take the equation of the hyperbola as described in the space and time coordinates of S, and find the slope at any point:

$$x^2 - (ct)^2 = 1$$

Therefore,

$$2x - 2(ct)\frac{d(ct)}{dx} = 0$$

or

$$\frac{d(ct)}{dx} = \frac{x}{ct} \tag{4–7}$$

Now evaluate the slope at the point where the axis of x' cuts the hyperbola. This x' axis is the line $t' = 0$. But by the Lorentz transformations, we have

$$t' = \gamma(t - vx/c^2)$$

Hence at this point we have

$$t = vx/c^2$$

and so, from Eq. (4–7), we have (at B')

$$\frac{d(ct)}{dx} = \frac{x}{ct} = \frac{x}{vx/c} = \frac{c}{v} \tag{4–7a}$$

But the ct' axis for the frame S' is the line $x' = 0$, and since, again from the Lorentz transformations, we have

$$x' = \gamma(x - vt)$$

the equation of the ct' axis in terms of the coordinates x and ct is

$$x - \frac{v}{c}(ct) = 0$$

It follows that the slope of this line is given by

$$\text{slope of } ct' \text{ axis (as measured in } S) = ct/x = c/v \tag{4–7b}$$

verifying that the construction analyzed above has the correct properties.

With this piece of analytic geometry out of the way, we can see that the world line AQ (of the right-hand end of the object at rest in S) cuts the x' axis of S' at a point A' which is less than unit distance (OB') from O. But, in an exactly equivalent way, the world line $B'R'$ (of the right-hand end of the object at rest in S') cuts the x axis of S at the point B, which is less than unit distance (OA) from O. This discussion should make it clear that the question "Does the Fitzgerald-Lorentz contraction *really* take place?" has no single, unequivocal answer from a relativistic point of view. The whole emphasis is on defining what actual observations we must make if we want to measure the length of some object that may be in motion relative to us. And the prescription is simply that we measure the positions of its

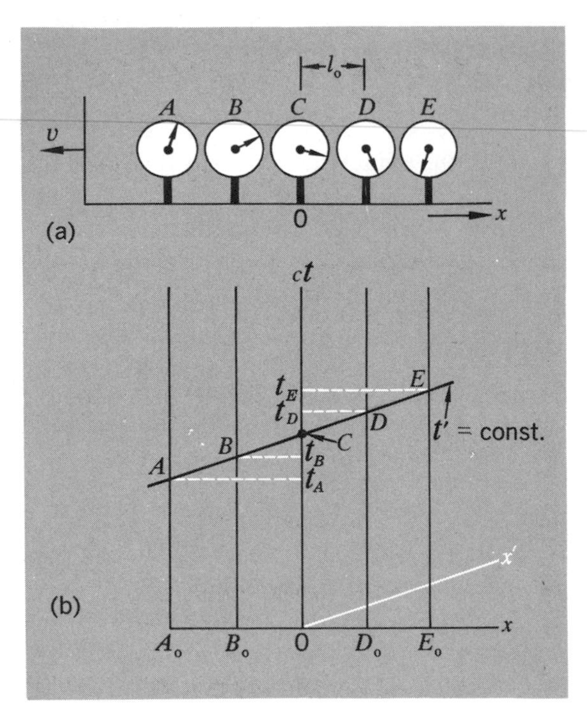

Fig. 4–7 (a) The readings, at a given instant in S′, of a row of clocks that are synchronized in S. (b) Space-time graph of this situation.

ends at the same instant as judged by us. What else could we possibly do?[1] Thus the contraction, when we observe it, is not a property of matter but something inherent in the measuring process.

The complete symmetry of the time dilation can be discussed in similar terms. For two frames in uniform relative motion, it is always a proper time interval, recorded by a single clock at rest in one frame, that is compared with an improper time interval as obtained from spatially separated clocks in the other frame. As a basis for discussing this a little more explicitly, suppose that we have five equally spaced clocks at rest in a frame S. The clocks are all synchronized according to the S definition of simultaneity. But now, suppose this whole system (S) is moving to the left with speed v as viewed from another frame (S'). Then according to the S' criterion of simultaneity, at any given instant (t') as measured in S' the clocks show a progressive difference of readings, as indicated in Fig. 4–7(a). This situation is depicted in different terms by means of the space-time graph in

[1]Other than the equivalent operation of noting the times at which its two ends pass a single fixed point in our frame, given a knowledge of its velocity relative to our frame.

Fig. 4–7(b). The vertical lines are the world lines of the clocks, and the points marked A, B, C, D, and E are the space-time coordinates of the clocks at a given time t' in S', specified by a line drawn parallel to the x' axis of S'. By the Lorentz transformations, we have

$$t' = \gamma(t - vx/c^2)$$

If the clock C, at the point $x = 0$ in S, reads t_0 at the time t', then we have

$$t' = \gamma t_0$$

and, substituting this particular value of t' back into the Lorentz-transformation equation, we see that the reading of a clock in S at any other value of x is given by

$$t(x) = t_0 + vx/c^2 \qquad (4\text{–}8)$$

This progressive "error" (as it would be judged to be from the standpoint of S') is, of course, very directly related to the radio-signal procedure for defining simultaneity. If a signal were sent out from C, then according to observers in S' the clocks A and B are moving away from it, whereas the clocks D and E are running to meet it. Being unaware of this, observers attached to S will set clocks too far ahead if they are at $x > 0$ and too far behind if they are at $x < 0$, exactly as stated by Eq. (4–8).

Now let us come to the time dilation itself and its symmetrical nature. Suppose we have *two* inertial frames, S and S', each of which has (according to its own measurements) a set of synchronized clocks spaced at equal intervals l_0 along the x axis. But suppose that one frame has the velocity $+v$, and the other the velocity $-v$, along x with respect to a third inertial frame, S''. Since the Lorentz contraction involves v^2 only, both sets of clocks appear equally spaced to S''. To simplify the discussion still further, we shall suppose that the progressive error, as judged by S'', between successive clocks in S or in S' is exactly 1 sec, as read on the clocks themselves; these errors run in opposite directions, of course, because of the oppositely directed velocities. Let a record be made of the clock readings of S and S' (as observed in S'') at those instants when their clocks are opposite each other. Figure 4–8 shows the possible results of three consecutive observations. If we imagine ourselves to be riding along with any one clock in either frame, we shall see that

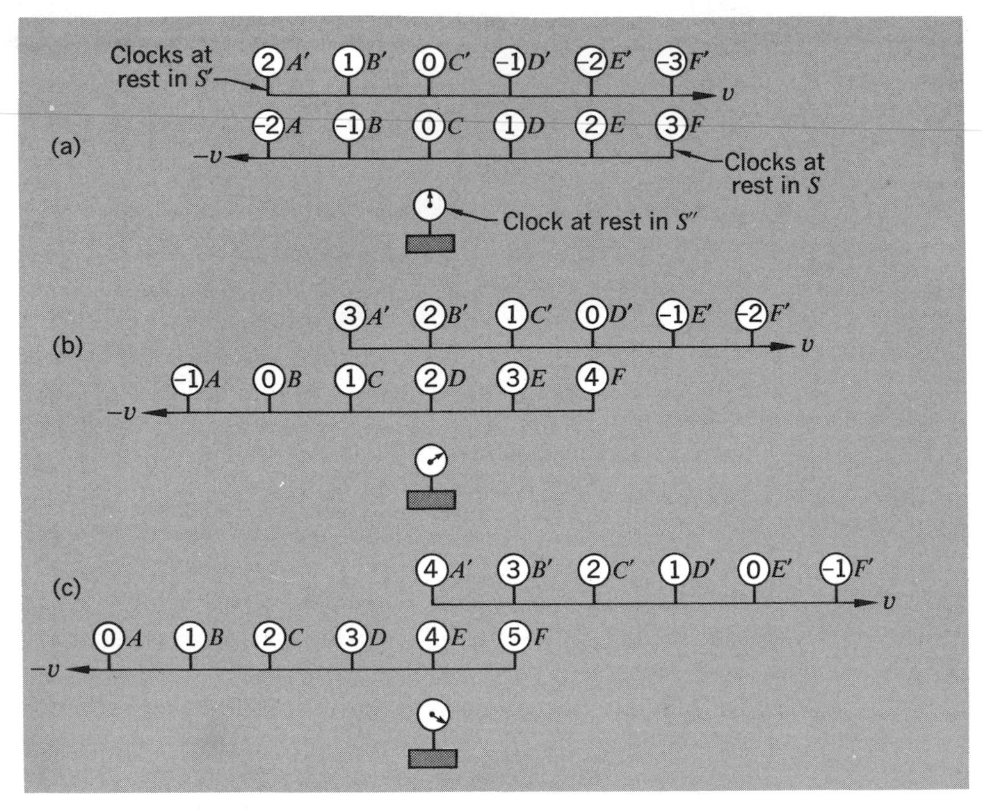

Fig. 4–8 Two rows of separately synchronized clocks moving with equal and opposite velocities with respect to a frame S''. Diagrams (a), (b), and (c) show readings of the clocks as they might be observed at three successive instants in S''.

it loses time steadily with respect to the adjacent clocks with which its readings are compared. For instance, the clock B' in S' is 2 sec ahead of B in (a), it agrees with D in (b), and it is 2 sec behind F in (c). But in an exactly similar way, each clock in S loses 2 sec with respect to clocks in S' at each stage [e.g., F is 6 sec ahead of F' in (a), 4 sec ahead of D' in (b), 2 sec ahead of B' in (c)]. There is no asymmetry provided we choose corresponding observations in the two frames. Relativity is truly relative.[1]

[1]This clear and pictorial argument is taken from W. Rindler, *Special Relativity*, Oliver and Boyd, Edinburgh, 1966. The particular values of clock readings postulated here require that $v = c/\sqrt{2}$ and that $l = \sqrt{2}c$ light-sec. See if you can prove this.

SPACE-TIME INTERVALS AND CAUSALITY[1]

In Chapter 3 we introduced the invariant quantity s^2 that can be constructed from the space and time coordinates of a given point event as measured in any one of a set of inertial frames having a common origin in space-time. The relationship, as described by Eq. (3–18), was as follows:

$$(ct')^2 - (x')^2 = (ct)^2 - (x)^2 = s^2$$

If we had considered two different events, we should find agreement among all inertial frames on the value of $(\Delta s)^2$, defined as the measure of $(c\,\Delta t)^2 - (\Delta x)^2$ in any frame. (And, since $y' = y$, $z' = z$, we could extend this equality to three-dimensional space without further ado.) The quantity Δs, dimensionally a distance, is called the *space-time interval* between two events; its value is zero for events connected via a light signal. Being an invariant Δs has a special importance, which can be illustrated with the help of space-time graphs. We shall not lay much stress on this space-time geometry, but a brief extension of our earlier discussion of such diagrams will perhaps be useful.

In Fig. 4–9 we show a space-time graph appropriate to one-dimensional problems. On it are marked two point events P and Q, together with the axes (x, ct) and (x', ct') appropriate to the different inertial frames. (Note that ct, rather than t, is once again used to describe the time variables, so that both coordinates have the dimension of distance.)

The two lines $x = \pm ct$ represent the world lines of light signals passing through the common origin O, and they divide the complete space-time into distinct regions. Those regions marked "elsewhere" cannot be reached by an observer whose world line passes through O, since to get into them from O would call for velocities greater than c. If the coordinate scales of x and ct are equal, the light lines are at 45° to the coordinate axes, as shown.

The quantity $(\Delta s)^2$ whose square root is the interval between two events P and Q can be positive, negative, or zero, according as $c\,\Delta t$ is greater than, less than, or equal to Δx. A light line connects point events between which the space-time interval is zero. If $c\,\Delta t > \Delta x$, as in Fig. 4–9, it is possible to transform to a reference

[1]This section can be omitted without loss of continuity.

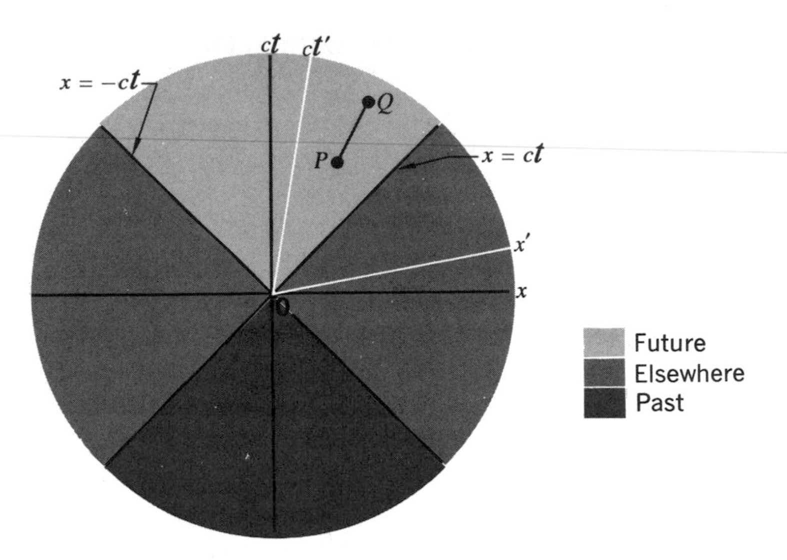

Fig. 4–9 Division of space-time (in a one-dimensional world) into past, future, and "elsewhere" from the standpoint of an observer at O. Also showing two events, P and Q, between which there could be a causal connection because the interval between them is "time-like."

frame whose axis of ct' is parallel to PQ. This means that P and Q are events occurring *at the same place* in S' and separated only by some pure time interval. The interval between P and Q is said to be *time-like* in such a case.

If $c\,\Delta t < \Delta x$, one can in a similar way transform to a frame in which P and Q are *simultaneous* events occurring at different places. The relation between them is called *space-like*. To connect two events that are separated by a space-like interval would require a velocity greater than c. Were this possible it would upset our ideas about causality in physics. For suppose an event $P(x, t)$ could cause an event $Q(x + \Delta x, t + \Delta t)$ through the agency of a signal having a velocity $u > c$. Then the time interval between the events as viewed in some other frame moving with velocity v would be given by

$$\begin{aligned} \Delta t' &= \gamma(\Delta t - v\,\Delta x/c^2) \\ &= \gamma\,\Delta t(1 - uv/c^2) \end{aligned} \tag{4–9}$$

Thus if $u > c$, we could find a range of values of v ($<c$) such that $\Delta t'$ and Δt were of opposite sign. All inertial observers with velocities greater than c^2/u could conclude from their observations that event P was caused by Q, rather than the other way round. This, however, would be very unpalatable, for it would make the laws of physics appear different to different observers—at least as long as we have a basis for knowing the direction in which time is advancing, as judged, for example, by our own ageing.

The above considerations do not, however, prevent us from having purely geometric velocities greater than c. For example, the spot produced on the moon by a laser beam directed from the earth could easily be made to sweep over the moon's surface at a speed much greater than that of light. The distance of the moon from the earth is about 3.8×10^8 m. Thus a laser searchlight being rotated on its pivot with the quite modest angular velocity of 1 rad/sec (one complete revolution in 2π sec) would already be enough to produce a spot speed of more than c. But no violation of dynamics is involved, for the path of the spot is nothing more than the locus of the points of impact of separate photons, each of which travels from earth to moon with speed c. The supervelocity is not associated with the motion of any material object or with the transport of energy or information from one point to another over the moon's surface.

If we consider a two-dimensional instead of a one-dimensional world, the space-time interval is defined by the equation

$$(\Delta s)^2 = (c\,\Delta t)^2 - (\Delta x)^2 - (\Delta y)^2 \tag{4–10}$$

The events which have a zero space-time interval with respect to the origin are thus given by

$$x^2 + y^2 = (ct)^2$$

This defines a surface that is called the *light cone*, part of which is indicated in Fig. 4–10. A figure such as Fig. 4–9 can then be regarded as a section of Fig. 4–10 taken in the xt plane. One can, of course, proceed further and write down an equation for the light cone in the case that all three directions in ordinary space are included. But if you want a picture of *this* light cone, you will have to construct it for yourself!

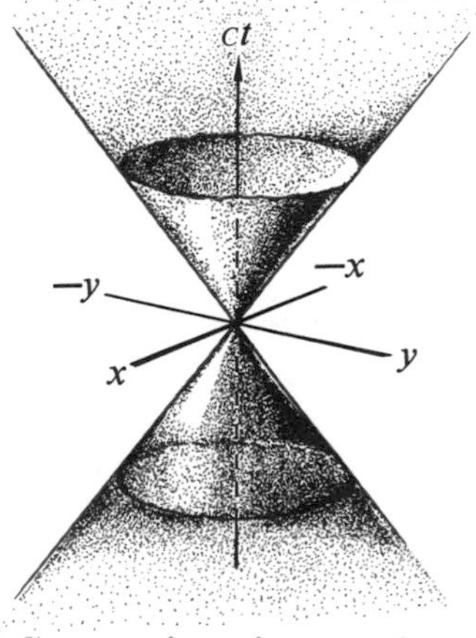

Fig. 4–10 Space-time diagram of a two-dimensional world, showing the light cone.

PROBLEMS

4–1 Two events occur at the same place in a certain inertial frame and are separated by a time interval of 4 sec. What is the spatial separation between these two events in an inertial frame in which the events are separated by a time interval of 6 sec?

4–2 Two events occur at the same time in an inertial frame S and are separated by a distance of 1 km along the x axis. What is the time difference between these two events as measured in a frame S' moving with constant velocity along x and in which their spatial separation is measured as 2 km?

4–3 As the text points out, an observer does not have a complete view of what is happening everywhere in his reference frame at a given instant; he is aware only of what is happening at his location at that instant. Suppose a meter stick pointing in the x direction moves along the x axis with speed $0.8c$, with its midpoint passing through the origin at $t = 0$. Assume an observer is situated at the point $x = 0$, $y = 1$ m.

(a) Where (in the observer's frame) are the end points of the meter stick at $t = 0$?

(b) When does the observer see the midpoint pass through the origin?

(c) Where do the end points appear to be at this time?

[See G. D. Scott and M. R. Viner, *Am. J. Phys.*, **33,** 534 (1965), for a more detailed discussion of this problem.]

4–4 Our galaxy is about 10^5 light-years across, and the most energetic particles known have an energy of about 10^{19} eV. How long would it take a proton with this energy to traverse the galaxy as measured in the rest frame of

(a) The galaxy?

(b) The particle?

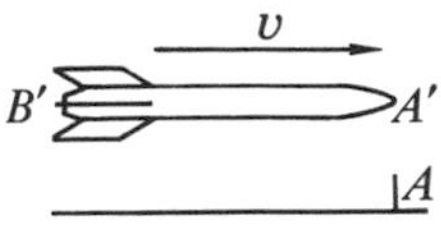

4–5 A rocketship of proper length l_0 travels at constant velocity v relative to a frame S (see the figure). The nose of the ship (A') passes the point A in S at $t = t' = 0$, and at this instant a light signal is sent out from A' to B'.

(a) When, by rocketship time (t'), does the signal reach the tail (B') of the ship?

(b) At what time t_1, as measured in S, does the signal reach the tail (B') of the ship?

(c) At what time t_2, as measured in S, does the tail of the ship (B') pass the point A?

4–6 A rocketship of length 100 m, traveling at $v/c = 0.6$, carries a radio receiver at its nose. A radio pulse is emitted from a stationary space station just as the tail of the rocket passes by.

(a) How far from the space station is the nose of the rocket at the instant of arrival of the radio signal at the nose?

(b) By space-station time, what is the time interval between the arrival of this signal and its emission from the station?

(c) What is the time interval according to measurements in the rest frame of the rocket?

4–7 An electron (rest energy 0.51 MeV) is accelerated from rest through a voltage drop of 0.13 MV, and then travels at constant velocity (1 MV $= 10^6$ volts).

(a) How long does it take for the electron (after it has reached its final velocity) to travel between two points 8.4 m apart?

(b) What is the distance between the points as measured in the rest frame of the electron?

4–8 A flash of light is emitted at position x_1 on the x axis and is absorbed at position $x_2 = x_1 + l$. In a reference frame moving with velocity $v = \beta c$ along the x axis:

(a) What is the spatial separation l' between the point of emission and the point of absorption of the light?

(b) How much time elapses between the emission and the absorption of the light?

4–9 Two spaceships, each measuring 100 m in its own rest frame, pass by each other traveling in opposite directions. Instruments on spaceship A determine that the front end of spaceship B requires 5.00×10^{-6} sec to traverse the full length of A.

(a) What is the relative velocity of the two spaceships?

(b) A clock in the front end of B reads exactly one o'clock as it passes by the front end of A. What will the clock read as it passes by the rear end of A?

4–10 Charged π mesons (pions) are produced in high-energy collisions between protons and neutrons. They decay in their own rest frame according to the law

$$N(t) = N_0(2)^{-t/T}$$

where T is the half-life and is equal to 2×10^{-8} sec. A burst of pions is produced at the target of an accelerator, and it is observed that two-thirds of them survive at a distance of 30 m from the target. What is the energy of the pions? (Express it in terms of the rest-mass energy.)

4–11 A beam of unstable K^+ mesons, traveling at a speed of $c\sqrt{3}/2$, passes through two counters 9 m apart. The particles suffer a negligible loss of speed and energy in passing through the counters but give electrical pulses that can be counted. It is observed that 1000 counts are recorded in the first counter and 250 in the second. Assuming that this whole decrease is due to decay of the particles in flight, what is their half-life as measured in their own rest frame?

4–12 At noon a rocketship passes the earth with a velocity $0.8c$. Observers on the ship and on earth agree that it is noon.

(a) At 12:30 P.M. as read by a rocketship clock, the ship passes an interplanetary navigational station that is fixed relative to the earth and whose clocks read earth time. What time is it at the station?

(b) How far from earth (in earth coordinates) is the station?

(c) At 12:30 P.M. rocketship time the ship reports by radio back to earth. When (by earth time) does the earth receive the signal?

(d) The station on earth replies immediately. When (by rocket time) is the reply received?

For maximum benefit, solve this problem from the standpoint of both the earth and the rocket frames.

(After J. H. Smith, *Introduction to Special Relativity*, Benjamin, New York, 1965.)

4–13 (a) In the time-dilation experiment with μ-mesons described in the text a layer of iron 2.5 ft high was used to slow down the mesons. It is nearly correct to say that μ-mesons traveling through iron suffer an average energy loss of $0.118 m_0 c^2/\text{cm}$, whatever their energy. If the energy lost by these particles in the scintillator itself is between zero and $0.45 m_0 c^2$, what was the initial velocity of the μ-mesons that stopped (and decayed) in the scintillator?

(b) If the μ-mesons all traveled in a direction making an angle θ with the vertical, what would be the ratio of the number of mesons decaying per hour in the scintillator on top of Mt. Washington to the number decaying in the scintillator at sea level? Show that if $\gamma \gg 1$, your answer is independent of θ. In the actual experiment it was assumed that $\theta = 0$ for all μ-mesons; estimate the error introduced by this assumption if in fact the detecting system accepts mesons at any angle up to $\theta = 45°$.

4–14 Two rockets are connected by an inextensible string of proper length l_0. At time $t = 0$ the rockets start out from rest with exactly equal constant accelerations as measured in S. At time $t = t_1$ the acceleration ceases and the rockets coast with equal constant velocities as measured in S. Why did the string break? [For further discussion see articles by E. Dewan and M. Beran, *Am. J. Phys.*, **27,** 517 (1959), and by E. Dewan, *Am. J. Phys.*, **31,** 383 (1963).]

4–15 A flash of light is emitted at point O and is later reabsorbed at point P (see the figure). In frame S, the line OP has a length l and makes an angle θ with the x axis. In a frame S' moving relative to S with a constant velocity v along the x axis:

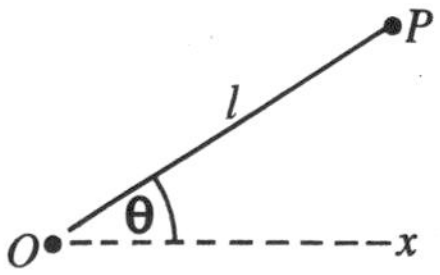

P
l
θ
O
x

(a) How much time τ' elapses between emission and absorption of the light?

(b) What is the spatial separation l' between the point of emission and the point of absorption of the light?

4–16 A rod of proper length l_0 is at rest in a frame S'. It lies in the (x', y') plane and makes an angle of $\sin^{-1}(\frac{3}{5})$ with the x' axis. If S' moves with constant velocity v parallel to the x axis of another frame S:

(a) What must be the value of v if, as measured in S, the rod is at 45° to the x axis?

(b) What is the length of the rod as measured in S under these conditions?

4–17 A rod of proper length L points along the x axis but moves in a direction making an angle of 45° to this axis (see the figure). A platform, also parallel to the x axis, lies in the rod's way, but a slit of proper length $1.1L$ has been cut out of it so that the rod can easily fit through if it travels at a nonrelativistic speed. What happens if its speed is $0.9c$? Analyze the problem from both reference frames.

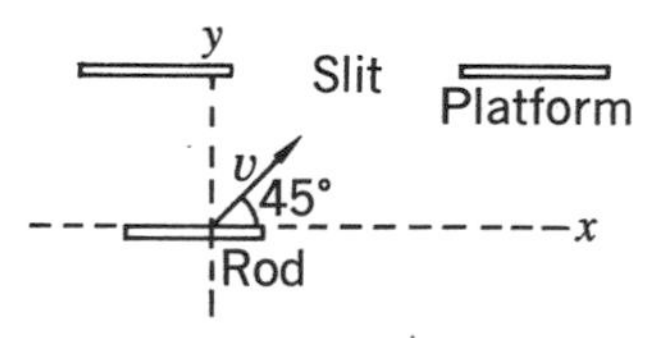

4–18 Below are the space and time coordinates of two pairs of events. Find the space-time interval between events 1 and 2 in each case. Taking each case in turn, answer the following questions:

(a) Could there be a causal connection between the two events?

(b) Is there a frame in which the two events would be recorded as simultaneous? If so, what is this frame?

	Event 1				*Event 2*			
	x_1	y_1	z_1	t_1	x_2	y_2	z_2	t_2
Case A	0.3 m	0.5 m	0	2×10^{-9} sec	0.4 m	0.7 m	0	3×10^{-9} sec
Case B	0.7 m	0.5 m	0	5×10^{-9} sec	0.4 m	0.6 m	0	4×10^{-9} sec

4–19 Consider a star S surrounded by a spherical shell of dust of radius R (see the figure). Assume that any light emitted by the star is first absorbed and then reradiated by the shell of dust. The star suddenly undergoes a nova-type explosion and sends out a very intense pulse of light. An observer on earth (very far away) sees light reradiated from A before he sees light coming from P. The total effect would be of an expanding ring of light with its center at A. Show that the rate (dr/dt) at which the radius of this ring appears to increase is given by

$$dr/dt = c \cot \theta$$

and so is greater than c for $\theta < 45°$. What about causality?

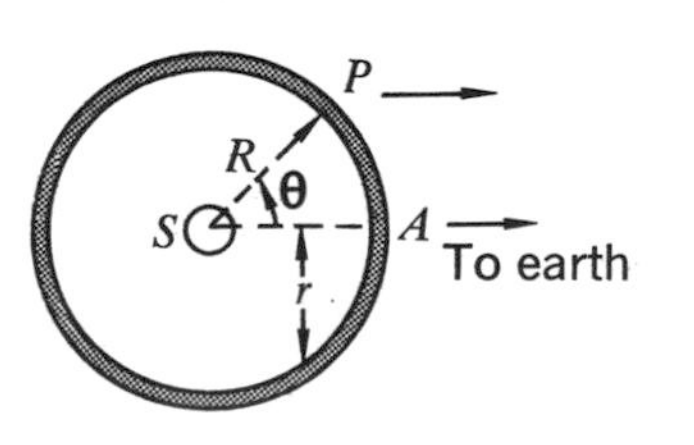

The velocity of light forms the upper limit of velocities for all material bodies . . . The simple mechanical law of adding and subtracting velocities is no longer valid or, more precisely, is only approximately valid for small velocities, but not for those near the velocity of light. The number expressing the velocity of light appears explicitly in the Lorentz transformation, and plays the role of a limiting case, like the infinite velocity in classical mechanics.

A. EINSTEIN AND L. INFELD,
The Evolution of Physics (1938)

5
Relativistic kinematics

TRANSFORMATION OF VELOCITIES

ONCE WE HAVE the Lorentz transformations for distance and time, it is a straightforward matter to consider the time derivatives of displacements as measured in two different inertial frames. Although we are in general concerned with motion in three-dimensional space, most of the results of interest can be developed in terms of motion in two dimensions. That is because (as we pointed out in Chapter 3) there is only one uniquely defined direction involved—the direction of relative motion of the inertial frames. This is an axis of symmetry, and any displacement can be analyzed into components along and transverse to it. Logically one should, perhaps, embody a reminder of this fact in the equations by using cylindrical coordinates, in which a vector is characterized by its component along the symmetry axis, its component transverse to this axis, and one angle—the azimuth—which is the angle between some fixed plane through the axis and the plane defined by the axis and the vector. In many instances the azimuth angle would then not appear in the results. Tradition has, however, enshrined the practice of using Cartesian coordinates, with y and z axes perpendicular to each other and to the direction (x) of relative motion of the reference frames. We shall therefore continue, as we have begun, with an acceptance of this coordinate representation. But the transformations appropriate to the x and y directions really tell the whole story and the transformations for the z direction can be constructed, usually by inspection, from those for the y direction. Therefore, unless there is special reason to do otherwise, we shall treat vectors as though they have x and y components only.

We shall begin, then, with the following basic equations:

$$\begin{aligned} x &= \gamma(x' + vt') \\ y &= y' \\ t &= \gamma(t' + vx'/c^2) \end{aligned} \tag{5-1}$$

with $\gamma(v) = (1 - v^2/c^2)^{-1/2}$.

Suppose that an object has velocity components u_x', u_y' as measured in S'. By the definition of velocity we have

$$u_x' = dx'/dt', \; u_y' = dy'/dt'.$$

What will be the relation between these and the components of the velocity of the object as measured in S, relative to which S' itself has the velocity v?

We have from equations (5–1), by differentiation,

$$\begin{aligned} dx &= \gamma(u_x' + v)\, dt' \\ dy &= u_y'\, dt' \\ dt &= \gamma(1 + vu_x'/c^2)\, dt' \end{aligned}$$

Hence

$$u_x = \frac{dx}{dt} = \frac{u_x' + v}{1 + vu_x'/c^2} \qquad u_x' = \frac{u_x - v}{1 - vu_x/c^2} \tag{5-2}$$

$$u_y = \frac{dy}{dt} = \frac{u_y'/\gamma}{1 + vu_x'/c^2} \qquad u_y' = \frac{u_y/\gamma}{1 - vu_x/c^2} \tag{5-3}$$

In Eqs. (5–2) and (5–3) we have added, for completeness, the expressions for u_x' and u_y' in terms of measurements made in S. Let us consider Eq. (5–2) first. It represents the relativistic law of addition for two velocities that are in the same direction. If both u_x' and v are small compared to c, the term vu_x'/c^2 is a quantity of the second order of smallness and can in most circumstances be ignored compared to unity. In this case we get back to the straightforward addition of velocities that works perfectly well for the kinematics of everyday life. But if either or both of the combining velocities should be comparable with c, we have results that are distinctively different from the Galilean kinematics.

Suppose, for example, that $u_x' = v = 0.5c$ so that we have $u_x' + v = c$. Then

$$u_x = \frac{0.5c + 0.5c}{1 + (0.5)^2} = \frac{4}{5}c$$

The combined velocity is less, by 20%, than we would have calculated by direct addition. Let us study this velocity-addition law in more generality by writing the combining velocities as fractions of c:

$$\begin{aligned} v &= \beta_1 c \\ u_x' &= \beta_2 c \end{aligned}$$

Then

$$\frac{u_x}{c} = \beta = \frac{\beta_1 + \beta_2}{1 + \beta_1\beta_2} \tag{5–4}$$

As long as the values of β_1 and β_2 are restricted to being less than unity (i.e., each of the combining velocities less than c) then the value of u_x/c is also less than unity. A simple way of seeing this is to consider the value of $1 - \beta$:

$$\begin{aligned} 1 - \beta &= 1 - \frac{\beta_1 + \beta_2}{1 + \beta_1\beta_2} \\ &= \frac{(1 - \beta_1)(1 - \beta_2)}{1 + \beta_1\beta_2} \end{aligned}$$

If β_1 and β_2 are both positive fractions, corresponding to the addition of velocities in the same direction, then the right side of the above equation clearly lies between 1 and 0; hence β lies between 0 and 1.

RADIATION FROM A RAPIDLY MOVING SOURCE

In the particular case that one of the combining velocities—say u_x'—is equal to c, giving $\beta_2 = 1$, then Eq. (5–4) yields the result $u_x = c$ for any value of v. This then includes the result that light emitted from a source that is moving relative to the laboratory still has the speed c, no matter how fast the source moves.

As we mentioned in Chapter 3, this result, although an essential feature of Einstein's formulation of special relativity, did not receive a convincing demonstration until much later. One crucial reason is that the propagation of light through a medium (even a transparent one) involves a continual process of absorption of the incident light and its reemission as secondary radiation by the medium—and it takes only a very small thickness of matter to bring about this replacement. Thus, for example, with visible light, a thickness of about 10^{-5} cm of glass or 0.1 mm of air at atmospheric pressure is almost enough to erase any possible memory, as it were, of the motion of the original source. This phenomenon, known as *extinction* (even though it may not involve any appreciable loss of intensity in the light beam), has invalidated some of the observations (e.g., the apparent motions of binary stars, already referred to in Chapter 3) that were at first believed to provide confirmation of Einstein's second postulate—the invariance of c. We shall now describe two experiments which do *not* appear to be vitiated by the extinction phenomenon.[1]

One experiment[2] (done, appropriately, at the Michelson Laboratory of the U.S. Navy) made use of visible light, in an apparatus very reminiscent of that for the Fizeau experiment (Chapter 2). Light from a primary source S (Fig. 5–1) passed via a mirror M to a beam splitter P so as to form two beams traveling in opposite directions around the path defined by mirrors $M_1 - M_4$. Two thin glass plates, P_1 and P_2, were mounted on a rotor pivoted at the point O so that P_1 and P_2 had equal and opposite velocities of magnitude v $(= \omega r)$. The primary source S was pulsed at the rotation frequency in such a way that no light circulated around the mirrors except during a few microseconds each revolution while P_1 and P_2 were in the positions shown. The whole apparatus was in a moderate vacuum (about 0.02 mm Hg).

If the light receives any velocity associated with motion of its source, then one would expect the modified speed to apply to the counterclockwise beam along P_1M_1 and P_2M_4, and to the clockwise beam along P_1M_2 and P_2P. Because of the extinction effect, any modification of speed must be eliminated in the process of reflection at any of the fixed mirrors. It follows that the modified speed, if it occurs, applies over a total distance l $(= M_1M_2 = M_4P)$ for each beam. If we suppose that a frac-

[1]Their validity has been carefully considered in an article entitled "Evidence against Emission Theories," by J. G. Fox, *Am. J. Phys.*, **33,** 1 (1965).

[2]G. C. Babcock and T. G. Bergman, *J. Opt. Soc. Am.*, **54,** 147 (1964).

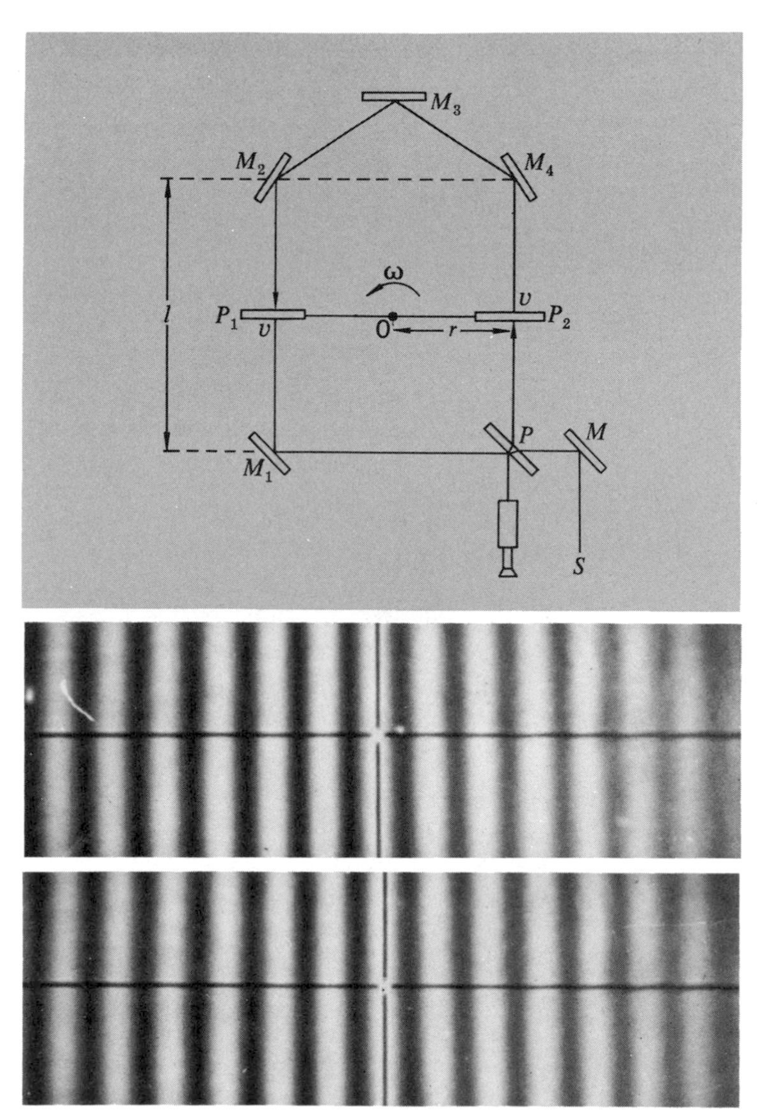

Fig. 5–1 (*a*) *Schematic diagram of apparatus of Babcock and Bergman for testing dependence of light velocity on source velocity. Can you suggest a possible reason why the extra mirror M_3 is present? (Consider the paths of light rays traveling along M_1M_2 and M_4P at different distances from the rotation axis O.)* (*b*) *Fringe patterns obtained (upper) with rotor stationary and (lower) rotating at 2800 rpm. No detectable shift occurs. (Photos courtesy of G. C. Babcock, Michelson Laboratory, Naval Weapons Center, China Lake, Calif.)*

tion f of v is given to the light, the time difference between the beams is given by

$$\Delta t = \frac{l}{c - fv} - \frac{l}{c + fv} \approx \frac{2lfv}{c^2}$$

The fringe shift corresponding to this would be $c\,\Delta t/\lambda$, and by reversing the direction of rotation one could double the effect. Hence we should have

$$\text{Fringe shift on reversal} = \frac{4lv}{\lambda c} f$$

In the experiment, the following values were used:

$$l = 2.76\text{ m}, \qquad \lambda = 4.74 \times 10^{-7}\text{ m},$$
$$v = 37.5\text{ m/sec } (\beta = 1.25 \times 10^{-7})$$

For $f = 1$ this would have given a shift of 2.9 fringes. Careful analysis of the actual fringe patterns revealed no significant shift as large as 0.02 fringe. Thus it could be concluded that the moving glass plates gave less than 1% of their velocity to the light leaving them.

An even more convincing test of the velocity-addition theorem when one of the combining velocities is c was that provided by the experiment we cited in Chapter 3, on the radiation emitted from neutral π mesons in flight.[1] The neutral π meson (or pion) may disintegrate into two photons:

$$\pi^0 \rightarrow \gamma_1 + \gamma_2$$

The pions were produced by the impact of protons of nearly 20 GeV energy on stationary nucleons in a target of a proton synchrotron.[2] This created pions with energies of more than 6 GeV, having $\gamma \geq 45$, $v/c \geq 0.99975$. (For a discussion of such particle creation processes, see Chapters 6 and 7.) These neutral pions have an extremely short mean life—only about 2×10^{-16} sec—so that even with the benefit of their large time-dilation factor, they travel an almost negligible distance (a few microns) before decaying. Decay photons emerging at about 6° to the proton beam were collimated, separated from charged particles by a magnet, and then timed very accurately over a flight path AB in air of about 30 m (see Fig. 5–2).[3] The timing was made possible by the fact that the protons in the accelerator came in very brief bursts, and so therefore did the decay photons. The time delay could then be measured between a pulse due to protons at the target and a pulse caused by the photons in the detector D. The change of this delay as the detector was moved from A to B could thus be recorded. The final result (already quoted in Chapter 3) was that the photon velocity was (2.9977 ±

[1] Alväger et al., *loc. cit.* [For a fuller account by the same authors, together with J. M. Bailey, see *Arkiv Fysik*, **31,** 145 (1966).]

[2] At CERN, the European nuclear research center in Geneva.

[3] The extinction effect was not important in this case, because for these very energetic photons the extinction length is about 5 km of air at atmospheric pressure.

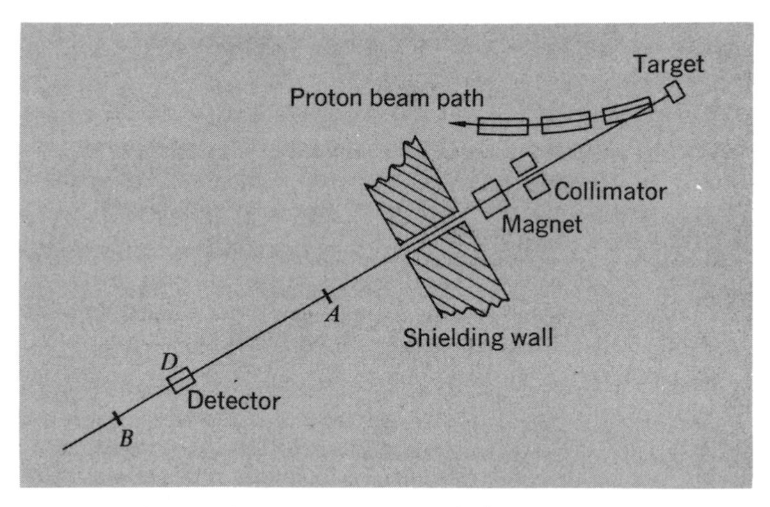

Fig. 5–2 Schematic diagram of experiment of Alväger et al. to measure speed of γ rays from decay of π^0 mesons.

0.0004) $\times$ 10^8 m/sec, which ruled out any augmentation of the photon velocity by more than about 10^{-4} of the velocity of the source (in this case the moving pions). Thus the relativistic law of velocity addition is upheld under very severe conditions, and the failure of the Galilean law is incontrovertible.

LIGHT IN A MOVING MEDIUM: THE DRAG COEFFICIENT

We have just considered two cases in which one of the two combining velocities was the velocity of light in free space. But now let us apply the relativistic velocity-addition law to the problem of light traveling within a moving medium, as in the Fizeau experiment. We shall then see that the famous *drag coefficient* emerges as a natural consequence of the new kinematics.

Here is how it works. The passage of light through a medium such as glass or water is characterized by a refractive index n, and the velocity of the light relative to the medium is c/n. Suppose that such a medium is moving with speed v parallel to the direction of the light. What light velocity V does a stationary observer find? Using Eq. (5–2) we have

$$V = \frac{c/n + v}{1 + v/nc}$$

Now for $v \ll c$ we can expand this expression in powers of v/c and ignore all powers higher than the first:

$$
\begin{aligned}
V &= \frac{c}{n}\left(1 + \frac{vn}{c}\right)\left(1 + \frac{v}{nc}\right)^{-1} \\
&= \frac{c}{n}\left(1 + \frac{vn}{c}\right)\left(1 - \frac{v}{nc} + \cdots\right) \\
&\approx \frac{c}{n}\left(1 + \frac{vn}{c} - \frac{v}{nc}\right)
\end{aligned}
$$

Therefore,

$$
V \approx \frac{c}{n} + \left(1 - \frac{1}{n^2}\right)v \tag{5–5}
$$

Thus the relativistic combination of velocities leads, without any extra assumptions, to precisely the result that Fresnel and the other ether theorists had to explain in terms of a partial dragging of the light by the medium. Prior to the development of Einstein's kinematics, the phenomenon was a mysterious one and the explanations of it were (as we have seen in Chapter 2) very artificial. To Einstein it was a famous and important result, which played a significant part in guiding him toward special relativity theory.

TRANSVERSE MOTIONS; STELLAR ABERRATION

Let us now consider the transformation of a velocity component perpendicular to the direction of relative motion of two reference frames. We have, by Eq. (5–3),

$$
u_y' = \frac{u_y/\gamma}{1 - vu_x/c^2}
$$

If we put $u_x = 0$ (a special case), we get

$$
u_y' = \frac{u_y}{\gamma} \qquad (\text{provided } u_x = 0) \tag{5–6}
$$

This apparent reduction of a transverse velocity component is essentially a manifestation of time dilation, as may be seen by considering the derivation of Eq. (5–3) from (5–1). There is really nothing more to be said. But if we consider the general form of the transverse velocity transformation, Eq. (5–3), and combine it with the longitudinal transformation, Eq. (5–2), we have a means of relating the *directions* of a given rectilinear motion as described in two different frames. One interesting

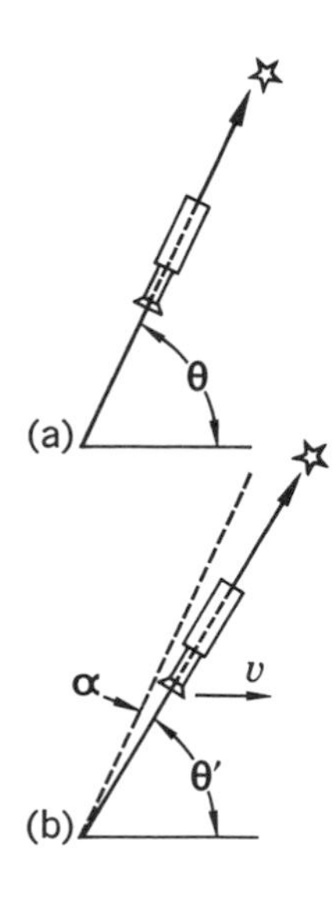

Fig. 5–3 Stellar aberration. (a) A stationary telescope aligned on a star. (b) A moving telescope aligned on the same star.

application of this result is to the problem of stellar aberration.

Regarding light as being composed of a rain of photons, we can easily calculate the change in apparent direction of a distant object such as a star. Let frame S be the rest frame of the sun, and let S' be the frame of the earth, traveling with the orbital velocity v relative to S (Fig. 5–3). Suppose the direction of a star, measured from the plane of the earth's orbit, is θ in frame S and θ' in S'. Then we have $u_x = -c\cos\theta$, $u_y = -c\sin\theta$ for the velocity components of the incoming photons as measured in S. Hence, using Eqs. (5–2) and (5–3), we have

$$u_x' = \frac{-(c\cos\theta + v)}{1 + v\cos\theta/c}$$

$$u_y' = \frac{-c\sin\theta}{\gamma(1 + v\cos\theta/c)} \tag{5–7}$$

The combination of the components u_x' and u_y' is, of course, a velocity of the invariant magnitude c. (Verify this for yourself.) Its direction θ' as observed in S' is most simply expressed by

$$\cos\theta' = \frac{-u_x'}{c} = \frac{\cos\theta + v/c}{1 + v\cos\theta/c}$$

As so often in relativistic calculations, it is convenient to denote the ratio v/c by β, so that we can write

$$\cos\theta' = \frac{\cos\theta + \beta}{1 + \beta\cos\theta} \tag{5–8}$$

Now, the earth in its motion around the sun carries us through a whole succession of reference frames, each of which can be regarded as approximately inertial during a short period of time. Since the orbital velocity of the earth is about 30 km/sec, the value of β in Eq. (5–8) is about 10^{-4}. An excellent approximation to Eq. (5–8) in this case is the following:

$$\cos\theta' \approx (\cos\theta + \beta)(1 - \beta\cos\theta)$$

i.e.,

$$\cos\theta' \approx \cos\theta + \beta\sin^2\theta \tag{5–9}$$

We introduce the angle of aberration, α, so that

$$\theta' = \theta - \alpha$$

Then

$$\cos\theta' = \cos\theta\cos\alpha + \sin\theta\sin\alpha$$

But α is a very small angle, so that we have, very nearly, $\cos\alpha = 1$ and $\sin\alpha = \alpha$, and hence

$$\cos\theta' \approx \cos\theta + \alpha\sin\theta \tag{5–10}$$

Comparing Eqs. (5–9) and (5–10), we see that

$$\alpha \approx \beta\sin\theta \tag{5–11}$$

which reproduces the approximate statement of the aberration as given in Chapter 2 in the discussion of Bradley's observations.

THE DOPPLER EFFECT

Some of the most important observations in atomic physics involve measurements on the radiation from moving atoms or nuclei. The apparent frequency (or quantum energy) of emitted radiation depends on the relative motion of source and observer. The term *Doppler effect* embraces all aspects of this phenomenon whereby the observed frequency or wavelength of a periodic disturbance is modified. The effect is named after Christian Doppler, who enunciated the essential principles of it in 1842, in connection with atomic spectroscopy.

Everyone is familiar with the acoustical Doppler effect—e.g., in the sudden drop in pitch of a siren on a car as the car passes by. In this case there is an identifiable medium—the air—and we have physically distinct situations according to whether it is the source or the observer (or both) that moves with respect to the medium. With light in vacuum there is no such distinction, so the discussion of the phenomenon is in a way simpler. It may be useful, nevertheless, to begin with a reminder of the acoustical Doppler effect.

Consider a one-dimensional problem in which a source S and a receiver R are moving along the same line (Fig. 5–4). Relative to the air, let the speeds of S and R be u_1 and u_2, respectively. Let the source be emitting a signal of frequency ν and period τ $(= 1/\nu)$. To make the discussion especially simple,

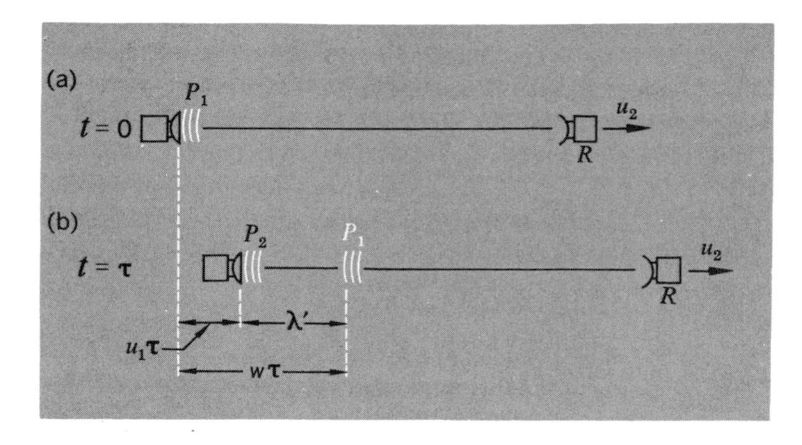

Fig. 5–4 Doppler effect for radiation from a moving source. (a) A first pulse leaves the source at $t = 0$. (b) A second pulse leaves the source at $t = \tau$. The motion of the source is such that the distance between pulses is less than if the source were stationary.

we shall assume that the signal is in the form of very brief pulses separated by τ. Each pulse travels through the air at the speed of sound, w. Suppose that a pulse P_1 is emitted at time $t = 0$, and a second pulse P_2 at $t = \tau$, as shown. During the time τ the pulse P_1 travels a distance $w\tau$ and the source S moves a distance $u_1\tau$. Thus the distance between P_1 and P_2, which we can call the effective wavelength λ' (although strictly this term should be reserved for pure sine waves), is given by

$$\lambda' = (w - u_1)\tau = \frac{w - u_1}{\nu}$$

The speed of the pulses relative to R is $w - u_2$, so that the time interval between the arrival of P_1 and P_2 at R is given by τ', where

$$\tau' = \frac{\lambda'}{w - u_2} = \frac{w - u_1}{\nu(w - u_2)}$$

The reciprocal of τ' defines an effective frequency ν'.[1] Thus we have the result

$$\nu' = \nu\frac{w - u_2}{w - u_1} = \nu\frac{1 - u_2/w}{1 - u_1/w} \tag{5–12}$$

For a given *relative* velocity v between R and S, the value of ν'

[1]The concept of frequency, like wavelength, applies strictly only to sine waves. Any other periodic signal with a repetition time τ has a frequency spectrum that in general may contain all harmonics of the frequency $1/\tau$.

still depends on the individual values of u_1 and u_2. Two cases of special interest are:

$$\left.\begin{array}{ll} \text{(a) Stationary source, moving receiver:} & \\ (u_1 = 0, u_2 = v) & \nu' = \nu(1 - v/w) = \nu(1 - \beta) \\ \text{(b) Moving source, stationary receiver:} & \\ (u_1 = -v, u_2 = 0) & \nu' = \dfrac{\nu}{1 + v/w} = \dfrac{\nu}{1 + \beta} \end{array}\right\} \quad (5\text{–}13)$$

The source and receiver are moving away from one another in both cases and the observed frequency is lower than the source frequency. Replacing v by $-v$ gives the corresponding results when source and receiver are approaching one another.

With this preliminary let us now turn to the corresponding problem in special relativity. We suppose that a source (e.g., a radar pulse transmitter) is located at the origin of reference frame S, and that an observer moves relative to S at velocity v, so that he is at rest in S'. Each emitted pulse travels with speed c. Suppose a first pulse is sent out at $t = 0$ when the observer is at the position $x = x_0$, and suppose the $(n + 1)$th pulse is sent out at $t = n\tau$. This will have covered n periods of vibration, so that the measured frequency of the source in S is $\nu = 1/\tau$.

To find what the observer records, we shall draw a graph of x versus t and identify the events representing the arrival of the first and last pulses at himself (see Fig. 5–5). In the language of such space-time diagrams, we locate the intersections of the world line of the observer with the world lines of the pulses. Let these intersections occur at (x_1, t_1) and (x_2, t_2) as measured in S. Then

$$x_1 = ct_1 = x_0 + vt_1$$
$$x_2 = c(t_2 - n\tau) = x_0 + vt_2$$

Therefore,

$$t_2 - t_1 = \frac{cn\tau}{c - v}$$
$$x_2 - x_1 = \frac{vcn\tau}{c - v}$$

But, as measured in S',

$$t_2' - t_1' = \gamma[(t_2 - t_1) - v(x_2 - x_1)/c^2] \qquad \text{(by the Lorentz transformation)}$$
$$= \gamma\left(\frac{cn\tau}{c - v} - \frac{v}{c^2}\frac{vcn\tau}{c - v}\right)$$

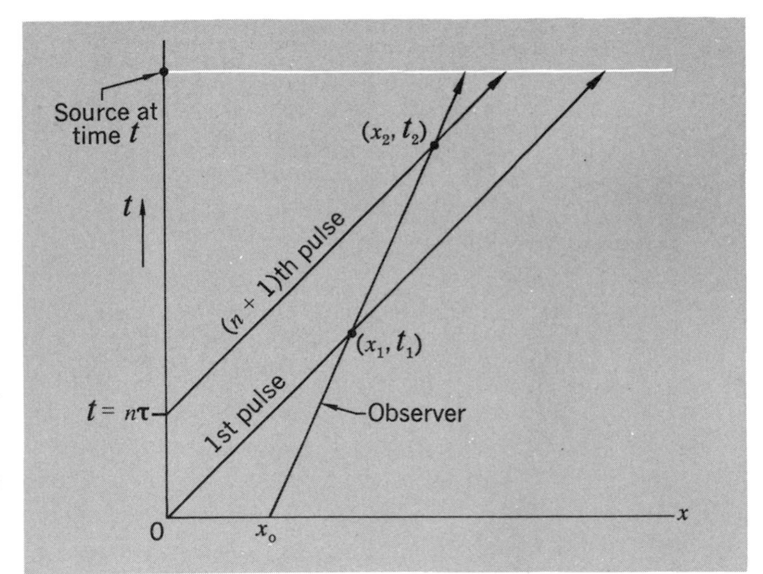

Fig. 5–5 Space-time diagram to illustrate the processes of emission and reception of light signals when source and receiver are in relative motion.

Since this time interval covers n periods of the signal as received by the observer, the apparent period τ' is given by

$$\tau' = \frac{\gamma c \tau}{c - v}\left(1 - \frac{v^2}{c^2}\right)$$

$$= \frac{\gamma(1 - \beta^2)}{1 - \beta}\tau \qquad \text{(putting } v/c = \beta\text{)}$$

$$= \gamma(1 + \beta)\tau$$

But

$$\gamma = (1 - \beta^2)^{-1/2}$$

Therefore,

$$\tau' = \left(\frac{1 + \beta}{1 - \beta}\right)^{1/2}\tau \tag{5–14}$$

or, in terms of frequencies,

$$\nu' = \left(\frac{1 - \beta}{1 + \beta}\right)^{1/2}\nu \tag{5–15}$$

Had S' been moving toward S, rather than away from S, the signs in numerator and denominator of the radical would have been interchanged. If you compare Eq. (5–15) with the acoustical Doppler formulas of Eq. (5–13), you will see that the relativistic result is a kind of unification of the moving-source and moving-observer results and can be set equal to either if terms higher than the first order are ignored. But the relativistic result has a special symmetry that the previous result lacks.

The most dramatic manifestation of this form of the Doppler effect, for relative motion of source and observer along the line joining them, is the famous *red shift* of distant galaxies. The spectrum of a complete galaxy, being a synthesis from all the different radiating objects in it, is close to being a continuous smear. But astrophysicists are able to distinguish a few very prominent dark lines—i.e., narrow gaps in the otherwise continuous spectrum—produced as the escaping radiation passes through cooler gases or vapors and undergoes selective absorption before leaving the galaxy. Two such lines in particular, the so-called H and K absorption lines of ionized calcium, can be distinguished even when all other characteristic features have been lost. (Ionized calcium atoms present an extraordinarily high cross section for light of these particular wavelengths.) They lie near the extreme violet end of the spectrum for a stationary source, but have been observed drastically shifted toward the region of longer wavelengths for certain very distant galaxies. A selection of galactic spectra with progressively increasing Doppler shifts is shown on pages 140 and 141. In each photograph the spectrum of the galaxy appears as a rather ill-defined horizontal streak, interrupted by the H and K absorption gaps. A line spectrum from a laboratory source is recorded above and below each galactic spectrum for purposes of comparison. In the last photograph, for example, the H and K absorption lines are found to be shifted to a wavelength of about 4750 Å, as compared to about 3940 Å for a stationary source. This is a very large increase of wavelength—nearly 25%. Using Eq. (5–14) we have

$$\lambda' = c\tau' = \left(\frac{1+\beta}{1-\beta}\right)^{1/2} c\tau = \left(\frac{1+\beta}{1-\beta}\right)^{1/2} \lambda$$

Therefore,

$$\beta = \frac{(\lambda'/\lambda)^2 - 1}{(\lambda'/\lambda)^2 + 1} \tag{5–16}$$

Putting $\lambda'/\lambda \approx 4750/3940 = 1.21$, we find

$$\beta = \frac{0.46}{2.46} \approx 0.2$$

Therefore,

$$v = 0.2c \approx 6 \times 10^7 \text{ m/sec}$$

It is by now a familiar story how the distinguished astronomer Edwin Hubble, who did so much to advance the study of the depths of space outside our own galaxy, established the existence of a linear relation between the velocity of recession and the distance for remote galaxies. Part (B) of the illustration on pages 140 and 141 shows the data of Part (A) plotted so as to exhibit this spectacular relationship, which is known as *Hubble's law.* The determination of the galactic distances is much less direct and definite than the measurement of the Doppler shifts and ultimately involves such profound questions as whether space on the grand scale is describable by Euclidean geometry. But this is beyond the scope of our immediate topic, and if you want further details you should hunt them up for yourself in a book on astronomy.[1] It must suffice here to lay the chief emphasis on the Doppler shifts themselves.

MORE ABOUT DOPPLER EFFECTS

As the first Sputnik sped around the earth it emitted a radio-frequency signal that was picked up by many tracking stations. Figure 5–6 shows one example of such observations.[2] When the satellite is very far away, approaching or receding, it gives maximum or minimum Doppler frequency-shifts corresponding to the one-dimensional problem we have been discussing. But the switch from augmented to diminished frequency is not instantaneous, as it would be if the moving object passed right through the position of the observer. Instead, it follows a smooth curve that can yield information about the altitude as well as the speed of the moving source. Let us analyze a situation of this kind.

In Fig. 5–7 we show the path of a satellite passing at a height h above an observation point O. We shall regard the path as being an approximation to a horizontal straight line, so that the satellite's position can be described by the following equations:

$$x = vt \qquad y = h$$

The time $t = 0$ marks the instant when the satellite is directly overhead.

We suppose the satellite to have a transmitter that sends out

[1] See, for example, F. Hoyle, *Frontiers of Astronomy*, Harper, New York, 1955, or his beautiful, more recent book, *Astronomy*, Doubleday, New York, 1962.

[2] R. R. Brown et al. (M.I.T. Lincoln Lab.), *Proc. IRE* **45,** 1552 (1957).

The red shift of distant galaxies

A. EVIDENCE OF AN EXPANDING UNIVERSE[a]

It is extremely difficult to photograph the spectra of distant galaxies. They are so faint that they cannot be observed visually, even with the largest telescopes. The correct aim of the telescope must therefore be calculated from the nebula's position in a photograph. Then the telescope has to be kept directed at the object for hours to photograph its spectrum which, in the end, may only measure 2 by 1 mm. The spectrum obtained is like that of an average star (spectral type G)—continuous, with absorption lines. But the absorption lines are very indistinct, forming as they do the average for all objects in the entire galaxy. By 1917 Slipher had managed to photograph the spectra of 15 spiral nebulae. He was surprised to find that the absorption lines in 13 of these spectra were displaced toward the red. This would suggest that the galaxies were moving away from us. From the shift observed it was possible to calculate that they were receding at 400 miles/sec (640 km/sec), on the average. In 1919 Hubble discovered that all external galaxies whose spectra had been photographed and distances determined are moving away from us at velocities proportional to their distances from us. This was once more very amply proved within the next few years by the observational material which Humason amassed using the Hooker and Hale telescopes. (Photographs from the Mount Wilson and Palomar Observatories.)

[a] *Comments from Ernst and De Vries, Atlas of the Universe, Thomas Nelson, London,* 1961.

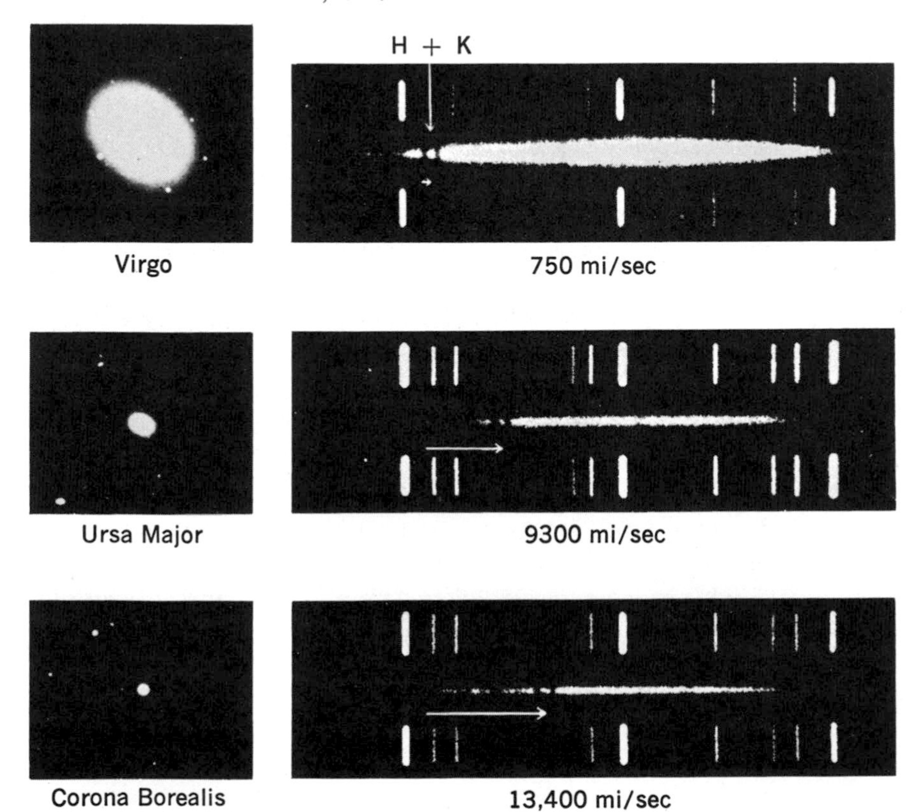

Comparison Lines of H and He

(wavelengths as marked)

3889 3970 4026 4100 4143 4340 4387 4471 4713 4860 4922 5016

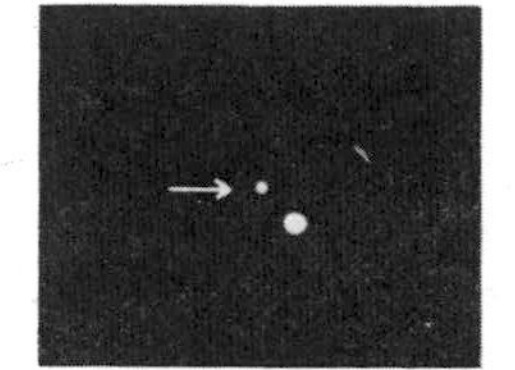

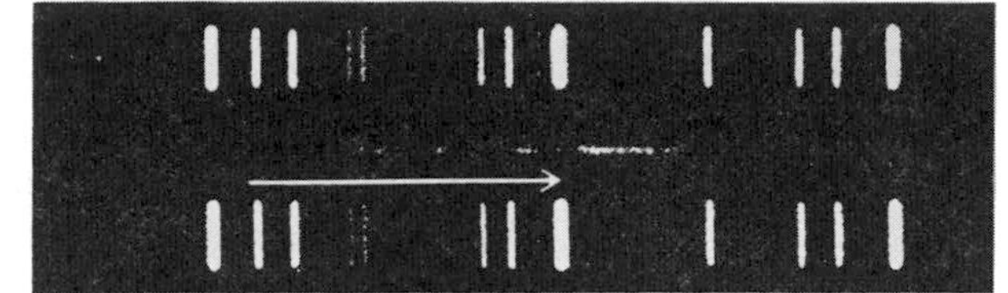

Boötes

24,400 mi/sec

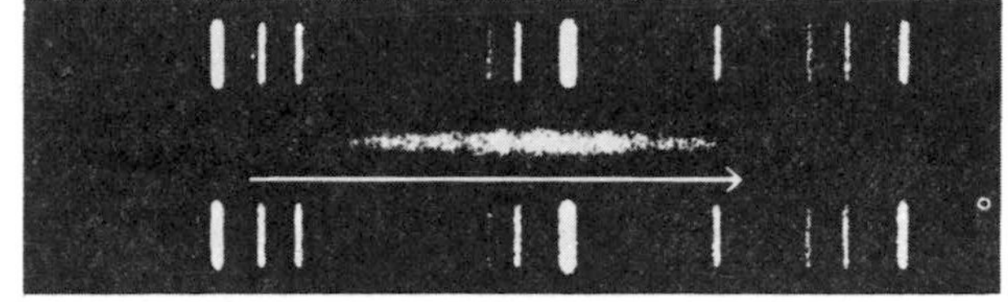

Hydra

38,000 mi/sec

B. HUBBLE'S LAW

The distances in the table are based on a 1958 revision of the distance scale. Velocities were calculated simply as $c\,\Delta\lambda/\lambda$, *with no account taken of relativistic modification of the Doppler formula.*

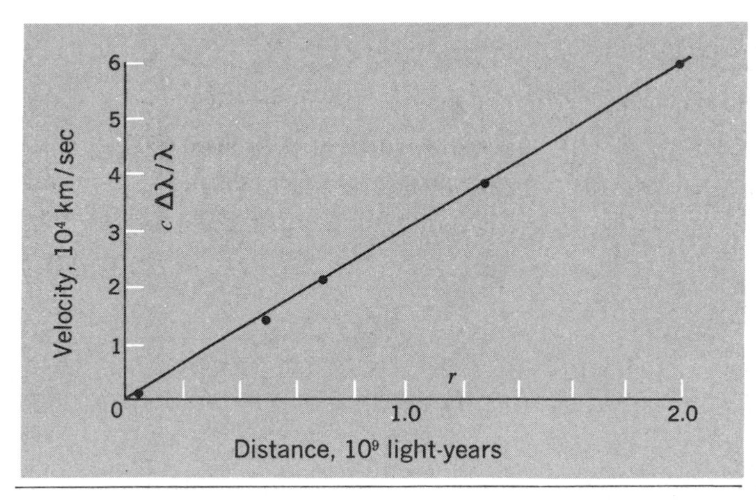

Galaxy in:	*Velocity,* $\times 10^4$ *km/sec*	*Distance, light-years*
Virgo	0.12	0.4×10^8
Ursa Major	1.40	5.0×10^8
Corona Borealis	2.14	7.0×10^8
Boötes	3.90	1.3×10^9
Hydra	6.10	2.0×10^9

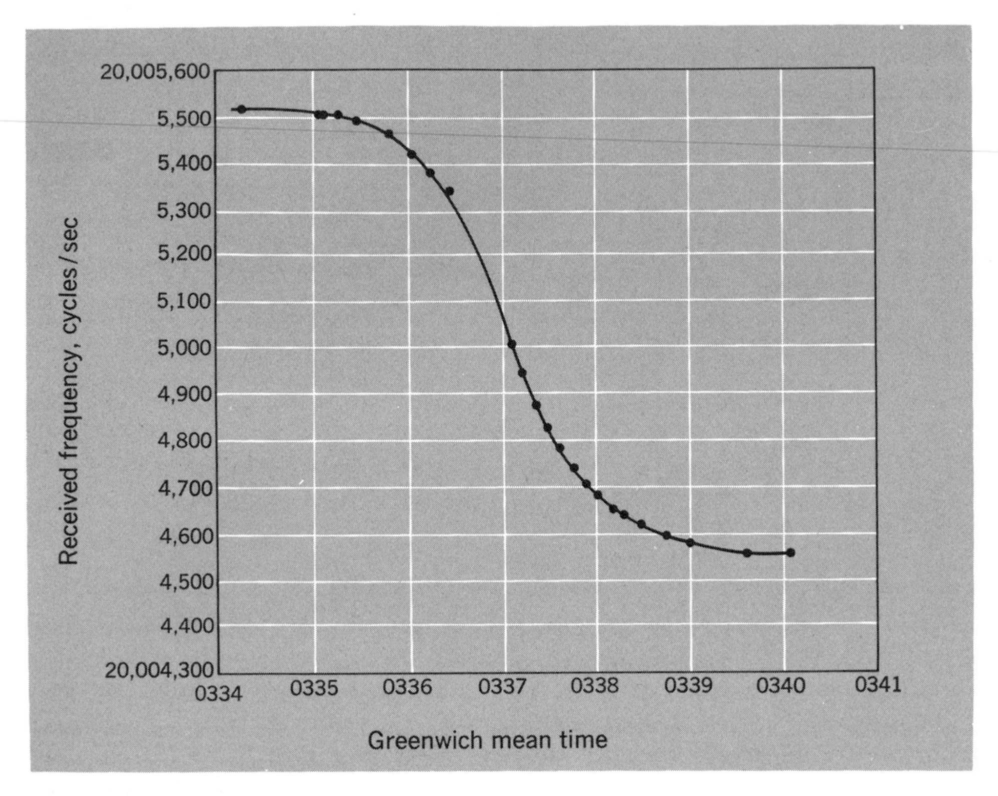

Fig. 5–6 Doppler effect of Sputnik I, observed at M.I.T. Lincoln Laboratory, Oct. 7, 1957. (Based on data of R. R. Brown et al.)

pulses at a frequency ν in its own rest system. Consider two successive pulses that are emitted from the positions x_1 and x_2 as shown, at times we can denote t_1 and t_2. The time interval τ between the pulses is $1/\nu$ in the inertial frame of the satellite but is greater than this by the time-dilation factor γ in the observer's frame. Thus we have

$$t_2 - t_1 = \gamma\tau = \gamma/\nu$$

The pulses take times r_1/c and r_2/c respectively to reach O, so that the measured time separation τ' between them is given by

$$\begin{aligned}\tau' &= t_2 + r_2/c - t_1 - r_1/c \\ &= \gamma\tau - (r_1 - r_2)/c\end{aligned}$$

Now if the distance $x_2 - x_1$ is very much less than r_1 (i.e., if the satellite travels a very small distance during one cycle of its transmitter signals), we can with good accuracy put

$$\begin{aligned}r_1 - r_2 &\approx (x_2 - x_1)\cos\theta \\ &= v\gamma\tau\cos\theta\end{aligned}$$

where θ is the angular elevation of the satellite at the instant t_1 as measured in the frame of reference attached to the ground. Thus we have

$$\tau' = \gamma\tau(1 - v\cos\theta/c)$$

But $1/\tau'$ represents the received frequency ν' of the signals under these conditions. Hence

$$\nu' = \frac{\nu}{\gamma(1 - \beta\cos\theta)}$$

i.e.,

$$\nu' = \nu\frac{(1 - \beta^2)^{1/2}}{1 - \beta\cos\theta} \tag{5-17}$$

If we wanted to proceed to construct (or analyze) the graph of observed frequency versus time (Fig. 5–6), we would make use of the relationship

$$\cos\theta = \frac{-vt}{(h^2 + v^2t^2)^{1/2}} \tag{5-18}$$

We did not really need special relativity to discuss the Doppler effect of Sputnik I, because the measurements involved were not sensitive to the differences of the order of β^2—i.e., a few parts in 10^{10}—between relativistic and nonrelativistic behavior in this case. It is true that, with the atomic clocks now available as frequency standards, such subtle changes are by no means beyond the reach of detection. But we shall not pursue this topic. The satellite problem simply provided a nice framework within which to develop the theory of Doppler effects for a source moving in an arbitrary direction. The really important applications of the Doppler formula as expressed by Eq. (5–17) are in the analysis of radiation from swiftly moving atoms, nuclei, or other subatomic particles. And as one example of this, we

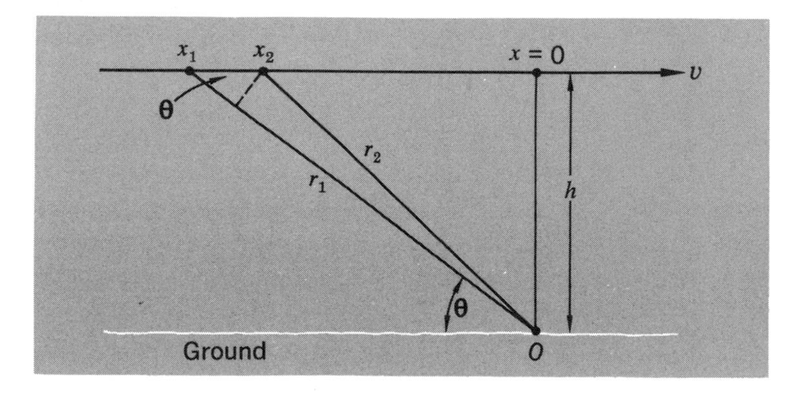

Fig. 5–7 Diagram for consideration of Doppler effect with signals emitted at angle θ to line of motion of source.

shall now discuss a classic experiment that exploited the Doppler effect to provide convincing quantitative evidence of the time-dilation phenomenon.

DOPPLER EFFECT AND TIME DILATION

Back in 1907 Einstein had suggested that a measurement might be made of the apparent wavelength of light emitted at right angles to their direction of motion by rapidly moving atoms. According to Eq. (5–17), the radiation traveling at an angle θ to the direction of a moving source has an observed frequency given by

$$\nu'(\theta) = \nu \frac{(1 - \beta^2)^{1/2}}{1 - \beta \cos \theta}$$

This defines an apparent wavelength given by

$$\lambda'(\theta) = \lambda \frac{1 - \beta \cos \theta}{(1 - \beta^2)^{1/2}} = \gamma\lambda(1 - \beta \cos \theta) \qquad (5\text{–}19)$$

The angle θ is the direction as measured by the observer. If we set $\theta = \pi/2$, the apparent wavelength is larger than λ by just the factor γ. Now if a proton accelerated through about 5 kV picks up an electron, it forms a hydrogen atom moving at a speed of about 10^6 m/sec, so that $\beta \approx 1/300$ and $\gamma - 1 \approx 5 \times 10^{-6}$. This value of $\gamma - 1$ represents the fractional change of measured wavelength for any light emitted sideways by the moving atom and for a line in the visible spectrum at 5000 Å would mean an absolute wavelength shift of about 0.025 Å. This is extremely small but might in principle be measurable. There is, however, a very serious practical difficulty. If one is to establish the existence of this transverse, or second-order, Doppler effect (as it is variously called), one must be sure that the angle θ is precisely $\pi/2$. A deviation from it by the amount β radians (equal to about 0.2° in this example) would cause the first-order Doppler factor (that is, $1 - \beta \cos \theta$) to swamp the effect being sought.

In 1938 H. E. Ives and G. R. Stilwell published the results of a beautiful experiment in which this difficulty was avoided.[1] Instead of trying to observe light emitted transversely to the direction of motion of the atoms, they made measurements on the radiation emitted forward or backward with respect to this

[1]H. E. Ives and G. R. Stilwell, *J. Opt. Soc. Am.*, **28**, 215–226 (1938).

direction. From Eq. (5–19) we have

$$\begin{aligned} \lambda'(\pi) &= \lambda \frac{1+\beta}{(1-\beta^2)^{1/2}} = \lambda \left(\frac{1+\beta}{1-\beta}\right)^{1/2} \\ \lambda'(0) &= \lambda \frac{1+\beta}{(1-\beta^2)^{1/2}} = \lambda \left(\frac{1-\beta}{1+\beta}\right)^{1/2} \end{aligned} \tag{5–20}$$

Let us develop these expressions in ascending powers of β:

$$\begin{aligned} \lambda'(\pi) &= \lambda(1 + \beta + \tfrac{1}{2}\beta^2 + \cdots) \\ \lambda'(0) &= \lambda(1 - \beta + \tfrac{1}{2}\beta^2 + \cdots) \end{aligned}$$

To the approximation that β^2 is negligible compared to 1, each of these wavelengths differs from λ by the first-order Doppler shift $\beta\lambda$—i.e., by about 15 Å for the hypothetical case we considered above. But when second-order effects are taken into account, one sees that the average of $\lambda'(\pi)$ and $\lambda'(0)$ differs from λ; it is, in fact, greater than λ by *precisely* the time-dilation factor γ, i.e., by a wavelength shift equal to $(\gamma - 1)\lambda$, as may be seen by direct substitution in Eq. (5–19). Making use of approximations where convenient, we can put

First-order effect:

$$\Delta\lambda_1 = \beta\lambda = \frac{v}{c}\lambda \tag{5.21a}$$

Second-order effect:

$$\Delta\lambda_2 = (\gamma - 1)\lambda \approx \frac{1}{2}\frac{v^2}{c^2}\lambda = \frac{1}{2\lambda}(\Delta\lambda_1)^2 \tag{5.21b}$$

In the absence of time dilation the value of $\Delta\lambda_2$ would be precisely zero [cf. Eq. (5–19) itself]; thus the measurement makes for a clear choice between the two versions of kinematics.

In the experiment, a hydrogen discharge tube was the source of ions of H_2^+ and H_3^+. (Free protons, H^+, were not observed in any appreciable amount, being quickly captured by hydrogen molecules to form the H_3^+ ions.) These ions, after acceleration through an accurately defined voltage, could (by neutralization plus dissociation) produce neutral but still excited hydrogen atoms whose velocities were in the ratio $\sqrt{3}:\sqrt{2}$, according to whether they came from H_2 or H_3. These atoms then emitted the characteristic Balmer lines of atomic hydrogen. Extremely careful observations were made, using a diffraction grating and

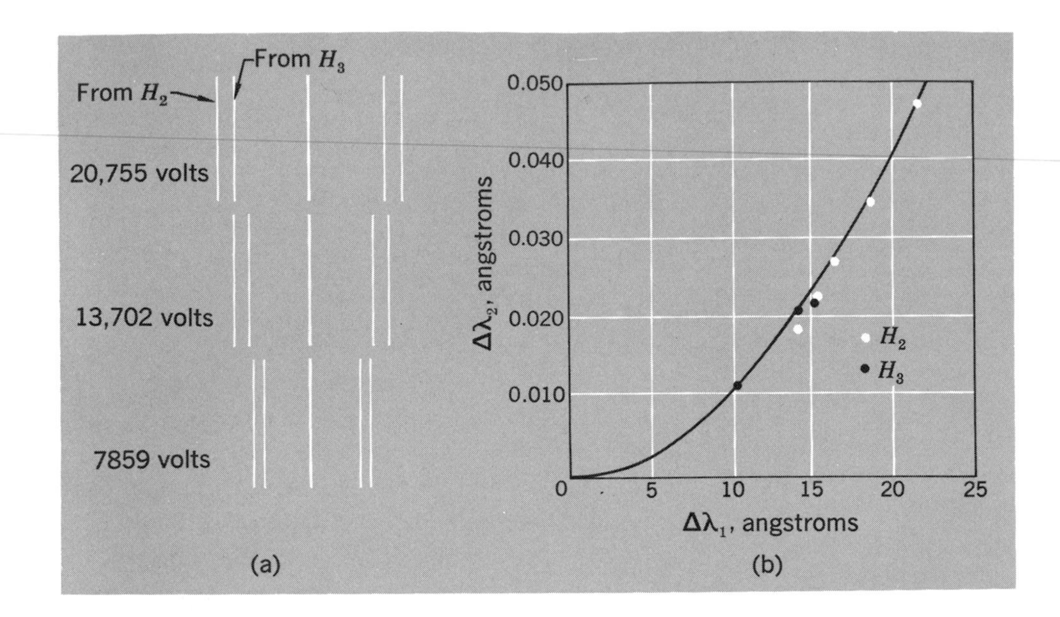

Fig. 5–8 Ives-Stilwell experiment (relativistic Doppler effect). (a) Spectrograms of a Balmer line (H_β) emitted forward or backward from H atoms traveling at different speeds. (b) Computed and observed second-order wavelength shifts, plotted against first-order Doppler shifts.

photographic recording, of the apparent wavelength of one particular line—the second line of the Balmer series, designated H_β. (It is turquoise blue and has a wavelength of about 4861 Å.) Figure 5–8(a) shows an example of the results obtained, with the unmodified line in the center and the Doppler-shifted lines for three different voltages on either side of it.

To a casual glance the Doppler shifts appear quite symmetrical with respect to the central line, but detailed measurement of the original photographs revealed a systematic displacement of the average wavelength with velocity as given by Eq. (5–21b). Figure 5–8(b) is a reproduction of the graph that Ives and Stilwell gave in their paper to show the parabolic relation between first-order and second-order shifts. The experiment, quite apart from the intrinsic interest and importance of its result, is an object lesson in what can be done by skilled investigators pushing a technique to its limit.

It is a curious sidelight on this experiment that its authors did not (even as late as 1938) accept special relativity theory. In their view the results simply demonstrated that a moving clock runs slow (as Larmor and Lorentz had suggested) by just the same factor, and in just as real a way, as a moving rod was believed to be contracted if it pointed along its direction of absolute motion through the ether. Old and cherished ideas die hard.

When Maxwell wrote his letter in 1879 that stimulated Michelson into looking for an ether drift, there was no conceivable way of doing a laboratory experiment that would reveal effects proportional to the first power of v/c. This was because any practicable experiment, such as Michelson's own, necessarily involved light rays turning back along their tracks, and as we have seen (Chapter 2) this reduced the observable effects to terms of the order of v^2/c^2 The development of the molecular-beam maser changed this and made possible a first-order experiment that placed far lower limits on v than had been achieved before.

The experiment, carried out by Cedarholm and Townes,[1] involved the comparison of the frequencies of two similar molecular-beam masers with their beams traveling in opposite directions. In very rough terms, the theory of the experiment from the standpoint of an ether theory is something like the following. The molecular-beam maser is somewhat like the light-pulse clock that we discussed in Chapter 4. It has a characteristic frequency defined by the time taken for radiation to bounce back and forth across it. (It is what is called a *resonant cavity*.) The radiation is supplied by a beam of molecules (excited ammonia molecules) traveling longitudinally relative to the cavity at some speed u [see Fig. 5–9(a)]. The dimensions of the cavity are adjusted so that

[1]J. P. Cedarholm and C. H. Townes, *Nature*, **184**, 1350–1351 (1959). See also J. P. Cedarholm et al., *Phys. Rev. Letters*, **1**, 342–343 (1958).

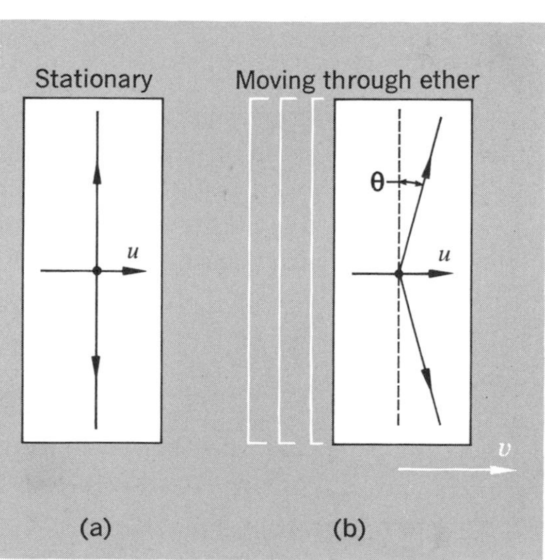

Fig. 5–9 Diagrams to illustrate emission of radiation from molecules in a molecular-beam maser that is assumed to be (a) at rest in the ether, (b) moving relative to the ether.

its natural frequency is about the same as that of the molecular radiation, but the really sharp definition of frequency is provided by the molecules, not by the cavity itself.[1]

Now if the cavity is at rest in the ether, the pulses of radiation travel exactly perpendicular to the direction of molecular motion, as in Fig. 5–9(a). If, however, the apparatus has a velocity v, parallel to the molecular beam, through the ether, then the radiation, to stay with the cavity, as it were, must be emitted at an angle $\pi/2 - \theta$ with respect to the molecular beam [Fig. 5–9(b)]. The size of the angle θ is v/c. This same picture applies to the light emitted along the transverse arm of the Michelson interferometer, if we view things from the ether frame rather than from the laboratory. But with the molecular-beam maser we now have a new effect. Since the radiation is provided by molecules moving with some velocity u relative to the cavity, these molecules act as sources having a velocity component $u\theta$ along the direction of the rays. This means a Doppler shift such that

$$\frac{\Delta\nu}{\nu} = \frac{u\theta}{c} = \frac{uv}{c^2} \tag{5–22}$$

The frequency shift is proportional to the first power of u. If, then, one takes two similar masers, set up with their axes parallel but with their molecular beams traveling in opposite directions (both, however, parallel to v), then the frequency of one is raised and that of the other is lowered. By mixing their outputs one can produce a beat frequency f given by

$$f = (2uv/c^2)\nu$$

Turning the whole system through 180° will exchange the roles of the two masers and hence give a net frequency shift equal to $(4uv/c^2)\nu$. The resonant frequency ν for an ammonia maser is about $2.4 \times 10^{10}\ \text{sec}^{-1}$, and u is a typical thermal velocity of about 600 m/sec. Thus we have

$$\begin{aligned} f &\approx \frac{1.2 \times 10^3 \times 2.4 \times 10^{10}}{9 \times 10^{16}}\, v\ \text{sec}^{-1} \\ &\approx 3 \times 10^{-4} v\ \text{sec}^{-1} \end{aligned}$$

where v is the velocity through the ether, in meters per second. Putting v equal to the earth's orbital velocity (3×10^4 m/sec)

[1]Note that this is just the opposite of the behavior of the He/Ne lasers used in the modernized Michelson-Morley experiment (cf. Chapter 4). For these it was the cavity dimensions, not the atoms, that defined the precise frequency.

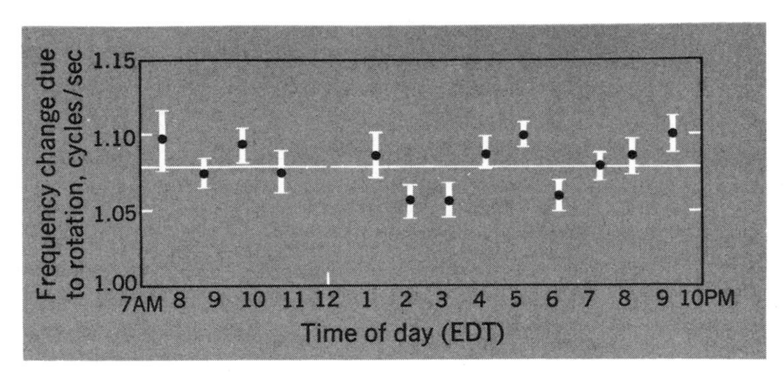

Fig. 5–10 Diurnal variation of the change in relative frequency caused by 180° rotation of a pair of ammonia beam masers (Cedarholm and Townes). The beams of the two masers were oppositely directed and in an east-west direction. The constant change of about 1.08 cps is primarily due to local magnetic fields. Lengths of lines indicate probable errors on the individual points.

would thus give a beat frequency of about 10 cps and a net beat frequency swing of 20 cps during a complete rotation of the apparatus. Figure 5–10 shows a sample of the actual results, as the apparatus was carried through 360° by the rotation of the earth. The measurements were repeated at intervals throughout a year, and at no time was a frequency shift of more than 0.02 cps detected. This puts an upper limit of about 30 m/sec (or about 1/1000 of the earth's orbital speed) on the value of v, and may well be regarded as delivering the coup de grâce to the fixed-ether hypothesis.

LOOKING AT MOVING CLOCKS AND OTHER OBJECTS

It may be appropriate here to set down a few remarks about the precise content of the statement that "moving clocks run slow." Whenever you see or hear this statement, you should conjure up in your mind a picture (artificial though it may be) of an inertial reference frame thickly sown with identical clocks that have all been synchronized by the radio-signal technique. As the moving clock travels through this territory, its reading at any particular point is compared with a stationary clock at that same point. From such observations it is concluded, as we saw in Chapter 4, that the moving clock is indeed running slow by the factor $(1 - v^2/c^2)^{1/2}$.

But what if we look at a moving clock—*really* look at it,

that is—for example, by watching it through a pair of field glasses? This is not at all the same thing. It means that we are standing at one particular point of a reference frame, and that what we see at the time t on our clock represents the reading on the moving clock at some earlier time $t - r/c$, where r was the distance of that clock from us at the earlier time. The clock can be thought of as sending out signals at equal intervals τ of its own proper time, which are not seen by us until later. But this is none other than the Doppler effect! At some instant, as measured by us, we see the distant clock through our binoculars and it shows a reading t. At a time τ' later, again as measured by us, we see the moving clock reading $t + \tau$. If the clock is moving along a straight line passing through our own position, the relation between τ' and τ is precisely that of the one-dimensional Doppler effect:

$$\tau' = \left(\frac{1 + \beta}{1 - \beta}\right)^{1/2} \tau \qquad (5\text{–}23)$$

If the clock is moving toward us, the value of β in this formula is negative, so that $\tau' < \tau$ and the moving clock will appear to be running fast, not slow. If the clock were a collection of moving atoms emitting a characteristic spectral line, we should see the light as blue-shifted.

The moral of this is simply that, as always in relativity, it is essential to be completely specific about what particular process or event is being described. The word "observe" must not be used loosely or uncritically with respect to events taking place at distant points. The words "see" and "look" must at once be recognized as involving the finite time of transit for light. It seems almost incredible in retrospect, but for over 50 years after Einstein's 1905 paper there was an unchallenged belief among physicists that the Lorentz contraction of a moving body could be seen or photographed. The difference between this kind of observation and the observations of which the Lorentz transformations are the expression was not critically considered until, in 1959, J. Terrell showed that the Lorentz contraction is not in general perceived as such by the eye.[1] When proper account is taken of the time for light to travel to a stationary observer's eye from different parts of a moving object, one recognizes that the

[1] J. Terrell, *Phys. Rev.*, **116,** 1041–1045 (1959). See also V. F. Weisskopf, *Phys. Today*, **13,** 24–27 (1960), and G. D. Scott and M. R. Viner, *Am. J. Phys.*, **33,** 534 (1965).

appropriate instants at which light must start out from various points of the body must be different, and hence are associated with different positions of the body as a whole. The general result of this is that the body appears distorted. There is, however, a special result of great interest. If the moving object is far enough away to subtend a very small solid angle (so that to some approximation the rays of light reaching the observer are all parallel), it develops that the object appears as it would at rest, but rotated. One would thus see a moving meter stick as foreshortened, but only to the extent that corresponded to its apparent rotation, without involving any contraction as such. We shall analyze this phenomenon for a rectangular object moving parallel to one of its edges.

Consider a rectangular board, of length L_0 and width W_0 as measured in its own rest frame, moving at speed v parallel to the edges of length L_0 [see Fig. 5–11(a)]. Let it be viewed in its own

Fig. 5–11 (a) A rectangular object moving at speed v parallel to x. (b) The apparent positions of the corners A, B, C as recorded at a given instant by a distant observer looking in the y direction in the plane of the object. (c) It is inferred that the object is rotated in its own plane, but not Lorentz-contracted.

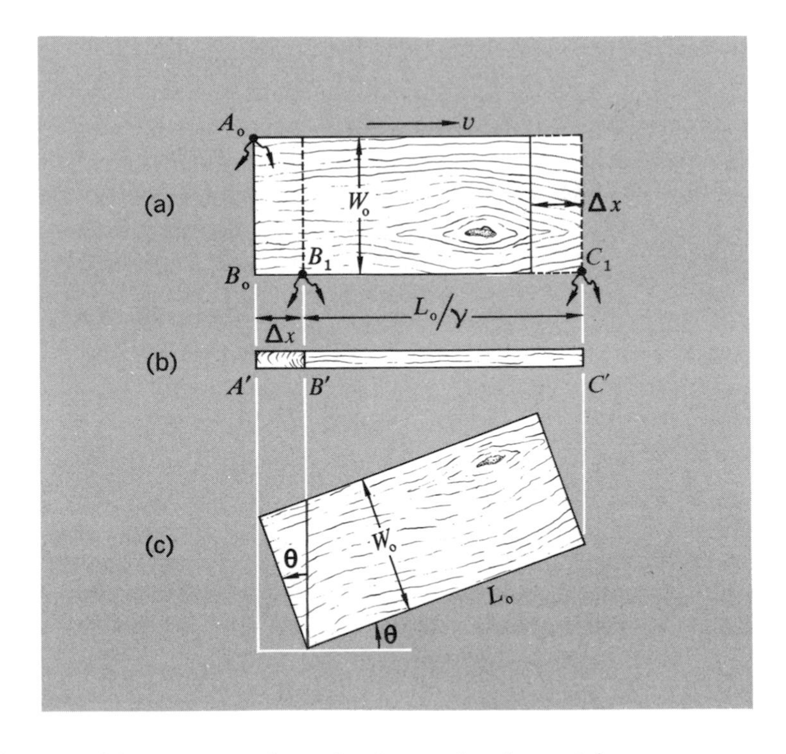

plane in a direction perpendicular to v. Then light from the rear corner (A_0) has to travel a distance W_0 more than that from the near corner (B_0), and thus, to arrive at the viewer's eye at the same instant, it must start out earlier by a time Δt equal to W_0/c. In this time, however, the board has moved a distance Δx equal to $v\,\Delta t$, that is, vW_0/c, and its ends are at the points marked B_1 and C_1 at the instant of departure of light from these ends. Because of the Lorentz contraction, the near edge of the board appears to be of length L_0/γ; since its ends B_1 and C_1 are equidistant from the observer, no correction for the difference of transmission times is involved. Thus, at a given instant, the observer gets a foreshortened view of the edge AB and a Lorentz-contracted view of the edge BC. This is indicated in Fig. 5–11(b), and we have the relationships

$$A'B' = vW_0/c \qquad B'C' = L_0(1 - v^2/c^2)^{1/2}$$

But now, if we consider the board at rest, but turned through an angle θ as shown in Fig. 5–11(c), the projected lengths of its edges are $W_0 \sin\theta$ and $L_0 \cos\theta$. These correspond precisely to the above values of $A'B'$ and $B'C'$ if we set $\sin\theta = v/c$.

Do not conclude from the above discussion that the Lorentz contraction is thereby liquidated—but realize that it does refer to measurements of a particular kind that are not the same as "just looking." This misconception, which must have made every physicist blush a little when it was pointed out, came directly out of a failure to analyze the seeing process in terms of the particular point events involved. Looking through your binoculars at a moving clock, you might very well in some circumstances see it running fast, but you would not necessarily see it as contracted. Be warned, therefore, and look with a cold and critical eye upon glib statements involving relativity theory.

ACCELERATED MOTIONS

For the sake of completeness, we shall develop here the expressions for accelerations as measured in different inertial frames. The results are somewhat complicated, and there is no point in trying to remember them unless relativistic kinematics is your livelihood. Even then, you would probably find special and simple cases more useful than the general results.

One other introductory comment is perhaps worth making.

Because Einstein developed a whole new theory (his general theory of relativity, published in 1916) based upon the dynamical equivalence of an accelerated laboratory and a laboratory in a gravitational field, it is sometimes stated or implied that special relativity is not competent to deal with accelerated motions. This is a misconception. We can meaningfully discuss a displacement and all its time derivatives within the context of the Lorentz transformations.

Just as with the velocity transformations, it is very advantageous to distinguish between longitudinal and transverse accelerations with respect to the direction of relative motion of two inertial frames. We have

$$u_x = \frac{u_x' + v}{1 + vu_x'/c^2} \tag{5-2}$$

$$u_y = \frac{u_y'/\gamma}{1 + vu_x'/c^2} \tag{5-3}$$

$$t = \gamma(t' + vx'/c^2) \tag{5-1}$$

Therefore,

$$du_x = \frac{du_x'}{1 + vu_x'/c^2} - \left[\frac{u_x' + v}{(1 + vu_x'/c^2)^2} \frac{v\,du_x'}{c^2}\right]$$

(Remember that v = constant for the purpose of this calculation.) Collecting the terms together, we have

$$du_x = \frac{(1 - v^2/c^2)\,du_x'}{(1 + vu_x'/c^2)^2} = \frac{du_x'}{\gamma^2(1 + vu_x'/c^2)^2}$$

Also, from Eq. (5–1),

$$dt = \gamma(dt' + v\,dx'/c^2) = \gamma(1 + vu_x'/c^2)\,dt'$$

Therefore,

$$a_x = \frac{du_x}{dt} = \frac{du_x'/dt'}{\gamma^3(1 + vu_x'/c^2)^3}$$

i.e.,

$$a_x = \frac{a_x'}{\gamma^3(1 + vu_x'/c^2)^3} \tag{5-24}$$

Similarly, from Eq. (5–3) we have

$$du_y = \frac{du_y'}{\gamma(1 + vu_x'/c^2)} - \frac{u_y'}{\gamma(1 + vu_x'/c^2)^2} \frac{v\,du_x'}{c^2}$$

Therefore,

$$a_y = \frac{du_y}{dt} = \frac{du_y'/dt'}{\gamma^2(1 + vu_x'/c^2)^2} - \frac{u_y'}{\gamma^2(1 + vu_x'/c^2)^3}\frac{v\,du_x'/dt'}{c^2}$$

i.e.,

$$a_y = \frac{a_y'}{\gamma^2(1 + vu_x'/c^2)^2} - \frac{(vu_y'/c^2)a_x'}{\gamma^2(1 + vu_x'/c^2)^3} \qquad (5\text{–}25)$$

Only if $u_y' = 0$ or $a_x' = 0$ (or both) does the expression for a_y become relatively simple. But for these cases we have

Special case ($u_y' = 0$ or $a_x' = 0$):

$$a_y = \frac{a_y'}{\gamma^2(1 + vu_x'/c^2)^2} \qquad (5\text{–}26)$$

It may be noted that if a body is instantaneously at rest in S' ($u_x' = u_y' = 0$), its acceleration components as measured in S are diminished by the factors γ^3 for the x direction and γ^2 for the y direction, as compared with the accelerations measured in the instantaneous rest frame S'.

The main lesson to be learned from the above calculations is that acceleration is a quantity of limited and questionable value in special relativity. Not only is it not an invariant, but the expressions for it are in general cumbersome, and moreover its different components transform in different ways. Certainly the proud position that it holds in Newtonian dynamics has no counterpart here.

THE TWINS

Of all the supposed paradoxes engendered by relativity theory, the *twin paradox* (or clock paradox) is the most famous and has been the most controversial. It asserts that if one clock remains at rest in an inertial frame, and another, initially agreeing with it, is taken off on any sort of path and finally brought back to its starting point, the second clock will have lost time as compared with the first. In today's parlance, the astronaut will end up by becoming younger than his twin brother. This result, which was stated by Einstein in his very first relativity paper (1905), became the subject of a raging controversy in the physics literature during the years 1957–1959, after preliminary skirmishes dating back to 1939. One must hope that the question has been finally settled.[1]

The "paradox" consists in a one-sidedness that appears to flout the basic tenets of relativity. Both the traveler and the stay-at-home agree that the traveler has aged less than the other. It is an entertaining application of the theory to see just why this should be so, and we shall give two similar but slightly different ways of analyzing the problem.

First, we define the traveler's journey in its simplest possible form, as composed of three distinctive events separated by periods of steady motion:

1. The traveler (A) starts off, reaching a constant velocity v within a negligibly short time.
2. After journeying for a while, the traveler suddenly reverses his velocity.
3. The traveler arrives back at his starting point, and stops.

Method 1. Suppose the observer on earth (B) records a total time T between events 1 and 3. He can infer that the outward and return journeys each took a time $T/2$. He (B) knows about time dilation, so he must be quite ready to concede that the traveler (A) records an elapsed time of only $T/2\gamma$ for each leg of the trip. Thus he should not be surprised that A measures the total journey time as T/γ. We can reinforce this conclusion through the Lorentz contraction. Event 2 occurs at a distance L ($= vT/2$) from the starting point. But A, just as soon as his acceleration up to v is completed, is in a frame in which the distance is only L/γ. Again we have a reduced value for the flight time. Both these calculations are correct.

But is it not possible to regard B as the traveler and A as the stay-at-home? No! Why not? Because there isn't symmetry between the two. Event 2 is the decisive one. During it, A switches from one inertial frame to another, while nothing at all happens to B. At his turnaround A experiences an acceleration that he can detect by various means. (Of course, if A is to start and stop at B's position, he undergoes accelerations at events 1 and 3 also.) He feels inertial forces acting on him. He sees stars! Moreover, if he has a telescope, he can see a sudden shift in the apparent position of those stars, owing to the v/c aberration

[1]See, for example, *Selected Reprints on Special Relativity Theory*, American Institute of Physics, New York, 1963; 9 of its 17 reprints are papers devoted to the clock paradox.

effects. There is no paradox, and the asymmetrical aging is real.[1]

Method 2. This method bases itself on the Doppler effect, and is perhaps even more convincing than method 1, of which it is, in effect, an elaboration. We imagine that each person sends equally spaced time signals (of his own proper time) to the other. The cumulative counts of time signals for the whole trip are then compared. Suppose each person is transmitting f pulses per unit time. As A travels away from B, each observer will receive the other's signals at the reduced rate $f' = f[(1 - \beta)/(1 + \beta)]^{1/2}$. But for how long? Here is the asymmetry. As soon as A reverses, he begins to receive signals from B at the enhanced rate $f'' = f[(1 + \beta)/(1 - \beta)]^{1/2}$. With B it is quite different. The last signal sent by A before he reverses does not reach B until a time L/c later. Thus for much more than one-half the total time B is recording A's signals at the lower rate f'. Only in the latter stages does B receive pulses at the higher rate f''. By drawing up a table (see Table 5–1) we can show that each observer receives as many signals as the other sends between start and finish of the trip. *They agree to disagree about their respective measures of the total time.* (It is worth noting in Table 5–1 that although B was not present at the space-time event representing A's turnaround, he is able to infer from his observations that it took place at the midmoment of the journey time as measured by A, since equal *numbers* of signals are received by B at the two different rates f' and f''.)

Even in the face of this analysis, it may still seem strange that two clocks, each of which has been reading proper time, should be brought together at the same spot and yet exhibit disagreement. After all, the separation in space-time between starting and finishing points is an invariant, can be verified by both observers, and is purely time-like. However, the summation of elementary space-time intervals ds along an actual path is, in Hermann Bondi's words, "a route-dependent quantity."

We have always

$$ds^2 = c^2\,dt^2 - dx^2 = (c^2\,dt'^2 - dx'^2)$$

[1]Note, though, that we *are* appealing to the reality of A's acceleration, and to the observability of the inertial forces associated with it. Would such effects as the twin paradox exist if the framework of fixed stars and distant galaxies were not there? Most physicists would say no. Our ultimate definition of an inertial frame may indeed be that it is a frame having zero acceleration with respect to the matter of the universe at large.

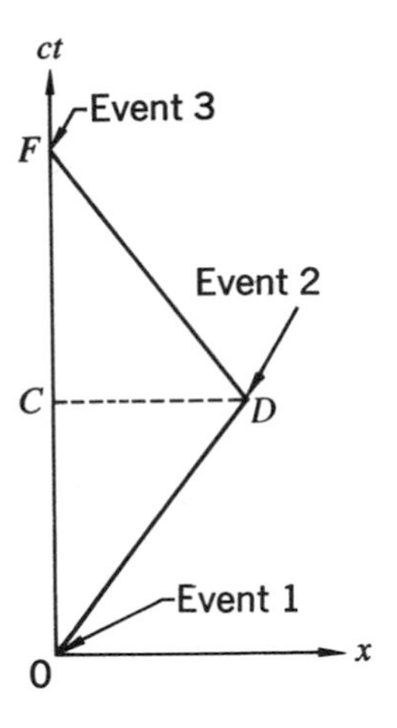

Fig. 5–12 Twin paradox. World lines of the astronaut (ODF) and his brother (OCF) in which the astronaut goes to a distant point and back again, traveling always at constant speed except at the start (O), turnaround (D), and finish (F).

But the integral of ds along the world line OCF of the stay-at-home (Fig. 5–12) is not the same as along the world-line ODF of the traveler (both as seen in the rest frame of B).

Along OCF, $dx = 0$ at every stage. Therefore,

$$\int_{1\ \text{path}\ OCF}^{3} ds = c\int_{1\ \text{path}\ OCF}^{3} dt = cT \tag{5–27}$$

Along OD, $dx = v\,dt$. Therefore,

$$\int_{1\ \text{path}\ OD}^{2} ds = (c^2 - v^2)^{1/2}\int_{1\ \text{path}\ OD}^{2} dt = cT/2\gamma$$

Along DF, $dx = -v\,dt$, so

$$\int_{2\ \text{path}\ DF}^{3} ds = cT/2\gamma$$

also. Hence

$$\int_{1\ \text{path}\ ODF}^{3} ds = cT/\gamma \tag{5–28}$$

We see that Eqs. (5–27) and (5–28) embody the by-now-familiar asymmetry. The result need not seem so very strange; it has an analogy in ordinary geometry. Just consider the case of two hikers, who decide to walk from a town P to a town Q. But one goes via village C and the other goes via village D. By consulting a map they agree on the value of the spatial interval between P and Q. But if they carry pedometers they will end up with different readings of the distance traversed in going from P to Q

TABLE 5-1: ANALYSIS OF THE TWIN PARADOX

Item	*Measured by B* (*the stay-at-home*)	*Measured by A* (*the traveler*)
Time of total trip	$T = 2L/v$	$T' = 2L/\gamma v$
Total number of signals sent	$fT = 2fL/v$	$fT' = 2fL/\gamma v$
Time of detecting *A*'s turnaround	$t_1 = \frac{L}{v} + \frac{L}{c} = \frac{L}{v}(1+\beta)$	$t_1' = \frac{L}{\gamma v}$
Number of signals received at the rate f'	$f't_1 = \frac{fL}{v}(1+\beta)\left(\frac{1-\beta}{1+\beta}\right)^{1/2}$	$f't_1' = \frac{fL}{v}(1-\beta^2)^{1/2}\left(\frac{1-\beta}{1+\beta}\right)^{1/2}$
	$= \frac{fL}{v}(1-\beta^2)^{1/2}$	$= \frac{fL}{v}(1-\beta)$
Time for rest of trip	$t_2 = \frac{L}{v} - \frac{L}{c}$	$t_2' = \frac{L}{\gamma v}$
Number of signals received at the rate f''	$f''t_2 = \frac{fL}{v}(1-\beta)\left(\frac{1+\beta}{1-\beta}\right)^{1/2}$	$f''t_2' = \frac{fL}{v}(1-\beta^2)^{1/2}\left(\frac{1+\beta}{1-\beta}\right)^{1/2}$
	$= \frac{fL}{v}(1-\beta^2)^{1/2}$	$= \frac{fL}{v}(1+\beta)$
Total number of signals received	$\frac{2fL}{v}(1-\beta^2)^{1/2} = \frac{2fL}{\gamma v}$	$\frac{2fL}{v}$
Conclusion as to the other man's measure of the time taken	$T' = 2L/\gamma v$	$T = 2L/v$

via C and D, respectively. The line integral of ds from P to Q depends on the path, and you would have a hard job to get a scientific wrangle going as to whether or why this should be so.

One last remark. It has been argued by some writers that an explanation of the twin paradox must involve the use of general relativity. The basis of this view is that the phenomena in an accelerated reference frame (including the behavior of a clock attached to such a frame) are regarded in general relativity as being indistinguishable, over a limited region of space, from the phenomena in a frame immersed in a gravitational field. This has been interpreted as meaning that it is impossible to talk about the behavior of accelerated clocks *without* using general relativity. Certainly the initial formulation of special relativity, although it leads to explicit statements about the rates of clocks moving at constant velocities, does not contain any obvious generalizations about accelerated clocks. And, as Bondi has remarked, not all accelerated clocks behave the same way. The clock consisting of a human pulse, for example, will certainly stop altogether if exposed to an acceleration of $1000g$—in fact, a mere $100g$ would probably be lethal—whereas a nuclear clock can stand an acceleration of $10^{16}g$ without exhibiting any change of rate. Nevertheless, for any clock that is not damaged by the acceleration, the effects of a trip can be calculated without bringing in the notions of equivalent gravitational fields. Special relativity is quite adequate to the job of predicting the time lost. It had better be, for (as Bondi has facetiously put it) "it is obvious that no theory denying the observability of acceleration could survive a car trip on a bumpy road." And special relativity has amply proved itself to be a more durable theory than this.

PROBLEMS

5-1 Consider three galaxies A, B, and C. An observer in A measures the velocities of C and B and finds they are moving in opposite directions each with a speed of $0.7c$ relative to him. Thus, according to measurements in his frame, the distance between them is increasing at the rate $1.4c$. What is the speed of A observed in B? What is the speed of C observed in B?

5-2 A K° meson at rest decays into a π^+ meson and a π^- meson, each having a speed of $0.85c$. If a K° meson traveling at a speed of $0.9c$ decays, what is the greatest speed that one of the π mesons can have? What is the least speed?

5–3 Consider two reference frames, S and S', moving with speed v ($< c$) with respect to one another along the x direction.

(a) If a certain object moves with velocity $\mathbf{u}$ ($u < c$) with respect to S, and velocity $\mathbf{u}'$ with respect to S', use the velocity addition equations (in three dimensions) to show that $u' < c$.

(b) If $u = c$, show that $u' = c$.

(c) If $v = 3c/4$, and $\mathbf{u}'$ has components $u_x' = -2c$, $u_y' = u_z' = 0$, show that the components of $\mathbf{u}$ are $u_x = 5c/2$, $u_y = u_z = 0$. How can you account for the fact that u_x and u_x' have opposite signs? Experiment (on paper) with some other examples in which either or both of the combining velocities is greater than c. Is there any physical significance to such situations?

5–4 Two neutrons, A and B, are approaching each other along a common straight line. Each has a constant speed βc as measured in the laboratory. Show that the total energy of neutron B, as observed in the rest frame of neutron A, is

$$(1 + \beta^2)(1 - \beta^2)^{-1}M_0c^2$$

where M_0 is the neutron rest mass.

5–5 A beam of atoms of radius R_1 moves with velocity u into a region of space containing a gas of atoms of radius R_2. If there are n such atoms per unit volume of the region, what fraction of the incident beam is scattered in time t measured in the laboratory frame? How does your answer change if the scatterers move with speed v (in the lab frame) into the incident beam? Assume that the density of scatterers is n in the scatterers' frame.

5–6 Measurements in two frames, S and S', are related by the usual Lorentz transformations, with $v = 0.6c$. At $t' = 10^{-7}$ sec, a particle leaves the point $x' = 10$ m, traveling in the x' direction with a constant velocity u' equal to $-c/3$. It is brought to rest suddenly at $t' = 3 \times 10^{-7}$ sec (all measurements in S'). As measured in S:

(a) What was the velocity of the particle during its trip?

(b) How far did it travel?

5–7 An inertial system S_1 has a constant velocity v_1 along the x axis relative to an inertial system S. Inertial system S_2 has a velocity v_2 relative to S_1. Two successive Lorentz-Einstein transformations enable us to go from (x, y, z, t) to $(x_1, y_1, z_1, t_1,)$ and then from $(x_1, y_1, z_1, t_1,)$ to (x_2, y_2, z_2, t_2). Show that this gives the same result as a single Lorentz-Einstein transformation from (x, y, z, t) to (x_2, y_2, z_2, t_2), provided we take the velocity v of S_2 relative to S as

$$v = \frac{v_1 + v_2}{1 + v_1v_2/c^2}$$

5–8 Consider two inertial frames, S and S', related in the usual manner.

(a) At $t = 0$ a photon leaves the origin of S, traveling in a direction making a 45° angle with the x axis. What angle does its trajectory make with the x' axis in S'?

(b) Repeat part (a) for a body of mass m moving in S with speed u.

(c) A rod which is stationary in S makes an angle of 45° with the x axis. What angle does it make with the x' axis?

5–9 Three identical radio transmitters A, B, and C, each transmitting at the frequency ν_0 in its own rest frame, are in motion as shown.

A B C
$-v$ $+v$

(a) What is the frequency of B's signals as received by C?

(b) What is the frequency of A's signals as received by C?

5–10 A pulsed radar source is at rest at the point $x = 0$. A large meteorite moves with constant velocity v toward the source; it is at the point $x = -l$ at $t = 0$. A first radar pulse is emitted by the source at $t = 0$, and a second pulse at $t = t_0$ ($t_0 < l/c$). The pulses are reflected by the meteorite and return to the source.

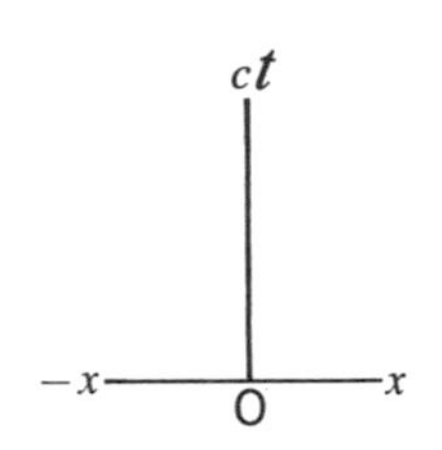

(a) Draw a coordinate system like that shown in the diagram, and on it mark position versus time (i.e., the world lines) for the following: (1) the source, (2) the meteorite, (3) the two outgoing pulses, (4) the reflected pulses.

(b) With or without the diagram, evaluate the time interval between the arrivals at $x = 0$ of the two reflected pulses.

(c) With or without the diagram, evaluate the time interval between the arrivals at the meteorite of the two outgoing pulses, as measured in the rest frame of the meteorite.

5–11 An astronaut moves radially away from the earth at a constant acceleration (as measured in the earth's reference frame) of 9.8 m/sec^2. How long will it be before the red shift makes the red glare of neon signs on earth invisible to his human eyesight?

5–12 There is a spaceship shuttle service from the earth to Mars. Each spaceship is equipped with two identical lights, one at the front and one at the rear. The spaceships normally travel at a speed v_0, relative to the earth, such that the headlight of a spaceship approaching earth appears green ($\lambda = 5000$ Å) and the taillight of a departing spaceship appears red ($\lambda = 6000$ Å).

(a) What is the value of v_0/c?

(b) One spaceship accelerates to overtake the spaceship ahead of it. At what speed must the overtaking spaceship travel (relative to earth) so that the taillight of the Mars-bound spaceship ahead of it looks like a headlight (5000 Å green)?

5–13 According to Hubble's law, the distant galaxies are receding from us at speeds proportional to their distance from us:

$$v(r) = \alpha r \qquad \text{with } \alpha \approx 2 \times 10^{-18}\ \text{sec}^{-1}$$

(a) How far away would a galaxy be whose speed with respect to the earth is c? Would it be observable from the earth?

(b) Consider the same questions (including Doppler effect) for a hypothetical galaxy for which $v(r) = 1.1c$.

5–14 A galaxy in Hydra emits light with a red shift corresponding to a recessional velocity with respect to the earth of 6×10^4 km/sec and, hence, according to Hubble's law, a distance of about 3×10^9 light-years. If this galaxy passed the earth T years ago and has moved with constant velocity ever since, what is the value of T?

5–15 Some observations reported on an astronomical object (quasar 3C-9) suggest that, when it emitted the light that has just reached the earth, it was moving away from the earth at a speed of about $0.8c$.

(a) One of the lines identified in its spectrum has a wavelength of about 1200 Å when emitted from a stationary source. At what wavelength must this line have appeared on the observed spectrum of the quasar?

(b) Quasars emit energy at such a huge rate that astronomers believe they must burn out in a relatively short time. If the lifetime of 3C-9 were assumed to be 10^6 years as measured in its own rest frame, over what total span of earth time would radiation from it be received at the earth? (Assume that its velocity relative to the earth remains constant.)

5–16 A calcium line in the spectrum of the star α Centauri has a measured wavelength of 3968.20 Å; the same line in the solar spectrum has a measured wavelength of 3968.49 Å.

(a) What is the radial velocity of α Centauri relative to the solar system? Is it approaching or receding?

(b) The transverse velocity of α Centauri is almost equal to its radial velocity; its distance from the sun is 4.3 light-years. By what angle does its apparent position in the sky change in 10 years?

(c) How fast would calcium ions need to be moving *transversely* to the line of sight if light of normal wavelength 3968.49 Å were to be changed by 0.29 Å? Would the light be distinguishable from that received from α Centauri?

5–17 (a) Consider two inertial frames, S and S', related in the usual manner. A light signal of frequency ν_0 in S is emitted from the point $x = -l$ at time $t = -l/c$. What is the frequency of the light signal in S'?

(b) Simplify your answer for the case $\beta c \ll 1$.

(c) Satisfy yourself that at $t = 0$, S' has the same velocity as a frame which began accelerating from rest at time $t = -l/c$ with acceleration $a = vc/l$.

(d) By the equivalence principle this result implies that the fractional frequency shift of light in a gravitational field g should be given by

$$|\Delta\nu|/\nu_0 = \beta = gl/c^2$$

where l is the distance between the emitter and the receiver. A classic experiment demonstrating this shift was performed by Pound and Rebka [*Phys. Rev. Letters*, **4,** 337 (1960)]. Their result was (measured shift)/(theoretical shift) $= 1.05 \pm 0.10$. In the experiment l was 60 ft; what was the value of $|\Delta\nu|/\nu_0$?

(e) Show that the fractional change in frequency of light (observed at infinity) leaving the surface of a star of mass M and radius R is $\Delta\nu/\nu_0 = -GM/Rc^2$.

5–18 A space traveler accelerates continually at a rate of 9.8 m/sec^2 in his instantaneous rest frame. If he starts at rest from the earth, how far has he traveled at earth time t? How long does it take before he attains a speed $c/2$? [You can use the transformation equation for the acceleration component a_x, Eq. (5–24), putting $u_x' = 0$ and recognizing that in this case $\gamma(v)$ is a function of time.]

5–19 The following problem is based on a delightfully simple and explicit numerical example of the twin paradox due to Sir Charles Darwin [Nature, **180,** 976 (1957)]. On New Year's Day, 1984, an astronaut (A) sets out from earth at speed $0.8c$ and travels to the nearest star, α-Centauri, which is just about 4 light-years away as measured in the earth frame of reference. Having reached the star he immediately turns around and returns to earth at the same speed, arriving home on New Year's Day, 1994, by earth time. The astronaut has a brother (B) who remains on earth, and they agree to send one another greetings by radar-telephone on every New Year's Day until the traveler gets back.

(a) Satisfy yourself that A sends only 6 messages (including the one made on the last day of his trip), whereas B sends 10.

(b) Draw a space-time diagram of A's journey as plotted in the earth reference frame. (Mark off the scales of both x and ct in light-years.) Draw also the world lines of all the radar signals that B transmits. Verify with the help of the diagram that A (the astronaut) has received only 1 signal up to the moment of his turnaround, and receives the other 9 during the return half of the trip.

(c) Draw another space-time diagram, again in the earth reference frame, showing the world lines of the astronaut and of all the signals that *he* sends. Verify that his brother receives one message each 3 years of earth time for the first 9 years after his brother's departure, and then receives 3 more messages during the last year, making a total of 6—which is just right, since for the astronaut the trip has taken 3 years each way.

(d) Interpret these results in terms of the Doppler effect.

5–20 A and B are twins. A goes on a trip to α Centauri (4 light-years away) and back again. He travels at speed $0.6c$ with respect to the earth both ways, and transmits a radio signal every 0.01 year in *his* frame. His twin B similarly sends a signal every 0.01 year in his own rest frame.

(a) How many signals emitted by A before he turns around does B receive?

(b) How many signals does A receive before he turns around?

(c) What is the total number of signals each twin receives from the other?

(d) Who is younger at the end of the trip, and by how much? Show that the twins both agree on this result.

5–21 An astronaut circles the earth at a radius of 7×10^6 m (from the center) for a week. How much younger than a twin remaining on earth is he when he lands? (Neglect the rotation of the earth.)

(*Note:* This is a case in which the traveler goes through an infinite number of different inertial frames, instead of just two, because of the constantly changing direction of his velocity. Basically, however, it comes down to a straightforward case of time dilation. There is a further complication in principle, associated with general relativity—that his body clock is in a different gravitational potential than at the earth's surface and this also affects the clock rate. But with the orbit radius so little different from the earth's radius, this effect is quite small compared to the time dilation.)

5–22 In the experiment designed by Alväger et al. described in the text, the decay of neutral π^0 mesons into γ rays (high-energy photons) was observed. These π^0 mesons had energies ranging (approximately) from 6.0 to 18.0 GeV, rest energies of 135.1 MeV, and proper half-lives of 2×10^{-16} sec.

(a) What was the range of speeds of the π^0 mesons?

(b) What was their range of half-lives (in the lab frame)?

(c) How far did they travel before decaying?

(d) The speed of the γ rays was determined by creating the π^0 mesons (and hence the γ rays) in bursts, and then measuring the time of flight of the γ rays in each burst moving a fixed distance in a chosen direction. If this distance was 60 m in the lab frame, what would the time of flight be for γ rays moving at speed c? At speed $2c$?

One can not insist sufficiently on the fact that the special theory of relativity today rests upon innumerable experimental verifications, for we can regularly obtain particles of velocities approaching that of light in vacuum, particles in regard to which it is necessary to take account of corrections introduced by the special theory of relativity.

L. DE BROGLIE (1949)

From an essay in *Albert Einstein: Philosopher-Scientist*, P. A. Schilpp, ed., a collection of essays published in honor of Einstein's 70th birthday, Harper Torchbooks, New York, 1959.

6

Relativistic dynamics—collisions and conservation laws

FACED WITH the results of special relativity, we should in principle rewrite all our mechanics accordingly. But we know that this is not necessary. The Newtonian scheme, although it is strictly correct only in the limit of vanishingly small velocities, works beautifully in an enormous variety of situations. This, as we have seen, is because the greatest velocities that we encounter in the dynamics of ordinary macroscopic objects are still minute compared to the velocity of light ($v < 10^{-5}c$). There is, however, one area in which the use of special relativity is clearly called for—in problems involving velocities that are *not* negligible compared with the velocity of light. And this means, primarily, the world of atomic and nuclear particles. It is with such problems, therefore, that this chapter will be largely concerned.

We shall not attempt in this chapter to consider in detail the motions of particles under the action of specified forces. Our goal will be a more modest one. What we shall do is to show the kinds of calculations one can do with the help of just two principles: (1) conservation of linear momentum, and (2) conservation of energy. No. (1) will, as in the familiar Newtonian problems, be applicable to each of the three separate components of linear momentum or to the total momentum treated as a single vector. No. (2) is used with the understanding that the total energy in all forms, including mass, is the conserved quantity. We shall take it as basic that these two principles apply to any self-contained system, and we shall concentrate on situations in which an interaction is over and done with in some limited span of time. In other words, we shall fix our attention on collisions or analogous processes, and our only concern will be to relate "before" and "after."

Before being able to apply these conservation principles, however, we must consider how to formulate and justify them in relativistic terms. In Chapter 1 we developed expressions for the mass, momentum, and total energy of a single particle of rest mass m_0 moving at speed v relative to the laboratory:

$$\begin{aligned} m(v) &= \gamma m_0 \\ \mathbf{p} &= \gamma m_0 \mathbf{v} \\ E &= \gamma m_0 c^2 \end{aligned} \tag{6-1}$$

with $\gamma(v) = (1 - v^2/c^2)^{-1/2}$. The derivation of these results made explicit use of the relation between energy and momentum for photons ($p = E/c$). Furthermore, if you consider the arguments in detail, you will see that, in fact, we *assumed* that conservation of momentum and energy held good—and then inferred the appropriate formulas for momentum and energy required by this assumption. Thus in considering the pressure of light experiment, we could not have inferred anything about the momentum of photons without assuming that this momentum was fully transferred to the illuminated surface. In the Einstein box calculation, also, the conservation of momentum is explicitly assumed. In discussing the ultimate speed experiment, we took it for granted that a calorimetric measurement would, through energy conservation, give us an exact knowledge of the kinetic energy brought in by electrons traveling at speeds close to the speed of light.

At this stage, therefore, we are not *discovering* these grand conservation principles of dynamics; we are, instead, *asserting* them, on the grounds that the use of such principles has already been amply justified in classical dynamics. And then, in going from Newtonian to Einsteinian dynamics, we are simply extending the range of problems that can be handled according to a single set of rules. In the process we arrive at new prescriptions for calculating such quantities as momentum and kinetic energy in terms of the velocity and an inertial parameter (the mass).

The transition to relativistic dynamics is not, however, an arbitrary one. Given the relativistic kinematics, and a knowledge of the Newtonian laws, one is led quite naturally to the relativistic formulation. We were not in a position to do this in Chapter 1, and our arguments there were, as we pointed out, more suggestive than convincing. But at this point we can offer a much cleaner approach, based upon readily visualizable situations which we shall now discuss. The course of the argument illustrates, once again, the very intimate connection between the particular formulation of kinematics and the dynamics appropriate to it.

TWO VIEWS OF AN ELASTIC COLLISION

We are going to consider a very simple type of collision process—a perfectly elastic collision between two identical particles. It will be a collision in which the whole motion takes place in one plane, and we shall analyze it in terms of momentum conservation.

By way of background, consider for a moment the Newtonian version of this process. Bodies A and B, with initial velocities $\mathbf{u}_1$ and $\mathbf{u}_2$, respectively, collide with one another and afterwards have velocities $\mathbf{v}_1$ and $\mathbf{v}_2$. In any individual collision of this type, it is always possible to find a set of four scalar multipliers (α) that permit one to write an equation of the form

$$\alpha_1\mathbf{u}_1 + \alpha_2\mathbf{u}_2 = \alpha_3\mathbf{v}_1 + \alpha_4\mathbf{v}_2$$

This as it stands is a quite uninteresting statement. But experiments for all sorts of values of $\mathbf{u}_1$ and $\mathbf{u}_2$ reveal the remarkable result that in every such collision, for two given objects, we can obtain a vector identity by putting $\alpha_1 = \alpha_3 = m_A$ (a scalar property of body A) and $\alpha_2 = \alpha_4 = m_B$ (the corresponding scalar property of body B). In other words, the purely kinematic observations on a collision process lead us to introduce the parameters that we call the *inertial masses* of the two bodies, and permit us to write the familiar equation for conservation of linear momentum.

When we introduce the relativistic kinematics, the relationship between initial and final velocities for two colliding objects is no longer expressible in quite such a simple form. Nevertheless, we keep as close to it as we can, and we do this by asking what is implied by the kinematics of such a collision if we assert conservation of linear momentum in the following extended sense.

In the elastic collision of two bodies, A and B, as described from the standpoint of a particular frame of reference, the initial and final velocities are related by the equation

$$m_A(u_1)\mathbf{u}_1 + m_B(u_2)\mathbf{u}_2 = m_A(v_1)\mathbf{v}_1 + m_B(v_2)\mathbf{v}_2 \qquad (6\text{–}2)$$

where m_A is a scalar inertial property of A depending only on its speed and m_B is the corresponding scalar property of B. We know that in Newtonian mechanics this equation is satisfied by values of m_A and m_B that are quite independent of speed. Let us now see how the relativistic kinematics implies the dependence of m on v as given in Eq. (6–1).

We imagine two experimenters, one in the inertial frame S and the other in S'. They use identical types of instruments for measuring times and distances, and they agree to produce a completely symmetrical collision between two identical particles.[1] The experimenter in S will project one particle (A) along his y axis with a speed u_0 (as measured in S), and the experimenter in S' will project the other particle (B) along his y' axis with a speed $-u_0$ (as measured in S'). *The speed u_0 is small, but S and S' have a very large relative velocity v along x.* The experimenters are so skillful that the particles collide when their centers lie along the y axis. The collision as observed in S and S' thus takes the forms shown in Fig. 6–1. The y (or y') component of velocity of each

[1] A gedanken experiment of this type was first introduced by G. N. Lewis and R. C. Tolman, *Phil. Mag.*, **18**, 510 (1909).

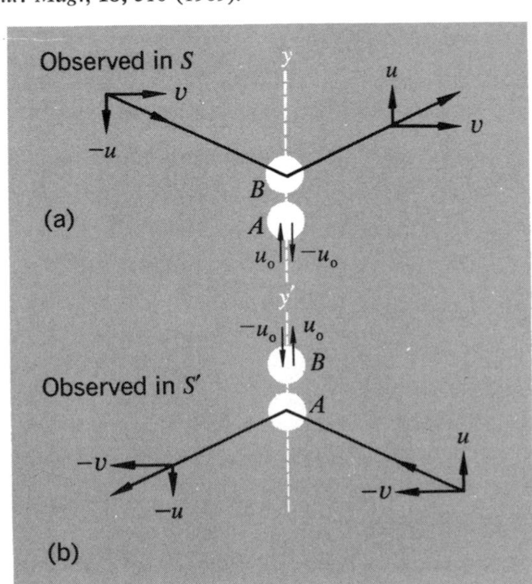

Fig. 6–1 Elastic collision between two identical objects, observed from two reference frames related by a velocity v along x.

particle is exactly reversed, and its velocity along x (or x') is unchanged. This corresponds to complete reversibility in time for the collision as a whole.

In analyzing this process, we note the following points:

1. As observed in S, the y component of velocity of A is initially u_0, and afterward $-u_0$. The y component of velocity of B is initially $-u$, and afterward u. The relation between u_0 and u is given by the transformation equation [Eq. (5–3)] for velocity components transverse to the direction of relative motion of two reference frames:

$$u_y = \frac{u_y'/\gamma}{1 + vu_x'/c^2}$$

Since in the frame S' the velocity component u_x' of B is zero, this reduces to

$$u = u_0/\gamma = u_0(1 - v^2/c^2)^{1/2} \tag{6–3}$$

2. As observed in S', the roles of A and B are interchanged and the sign of v is reversed. The complete symmetry can be clearly recognized if Fig. 6–1(b) is rotated through 180° in its own plane; it then matches Fig. 6–1(a) in every respect.

3. As observed in either reference frame, the *speed* of each particle remains unchanged by the collision, and is either u_0 or $(u^2 + v^2)^{1/2}$. Because of this, and the identity of the particles, we are concerned with only two possible values of m—$m(u_0)$ and $m(V)$, where $V = (u^2 + v^2)^{1/2}$.

The conservation of linear momentum in the y direction as observed in S is then described by the following statement:

$$p_y = m(u_0)u_0 - m(V)u = -m(u_0)u_0 + m(V)u$$

Therefore,

$$\frac{m(V)}{m(u_0)} = \frac{u_0}{u} \tag{6–4}$$

Now we have postulated that u_0 is small—as small as we choose to imagine. Hence the inertial quantity $m(u_0)$ can be taken to be just the Newtonian inertial mass m_0. Also, given that $u_0 \ll v$, it follows (a fortiori) from Eq. (6–3) that $u \ll v$, and hence that $V \approx v$. Thus, by imagining a limiting collision of this type, with $u_0 \to 0$, we conclude from Eq. (6–4) that

$$m(v) = \gamma m_0 = \frac{m_0}{(1 - v^2/c^2)^{1/2}} \tag{6–5}$$

and hence that $\mathbf{p} = \gamma m_0 \mathbf{v}$ is an appropriate definition of the linear momentum for a particle of rest mass m_0 traveling at velocity $\mathbf{v}$.

The above discussion is limited strictly to the question of momentum. Is the definition of mass that emerges from this analysis—i.e., as given by Eq. (6–5)—also applicable to calculations involving the *energy* of the system? We have of course already argued in Chapter 1 that this is indeed so, but an analysis based upon another hypothetical collision process will perhaps lend further conviction to the result.

TWO VIEWS OF AN INELASTIC COLLISION

Again we shall consider the impact of two identical particles, but this time we shall suppose that the collision is completely *in*elastic. There will be a frame S' in which the particles approach each other along a straight line with equal and opposite velocities of magnitude u [Fig. 6–2(a)]. There will then exist another frame S, relative to which S' has velocity u, in which one of the particles is initially stationary [Fig. 6–2(b)].

As observed in frame S', the collision results in the formation of a stationary composite particle. Hence in frame S this composite particle must be observed to have the velocity u. In this same frame S the initially moving particle has a velocity U related to u through the velocity addition formula, Eq. (5–2):

$$u_x = \frac{u_x' + v}{1 + vu_x'/c^2}$$

in which we put $u_x = U$, $u_x' = v = u$. Therefore,

$$U = \frac{2u}{1 + u^2/c^2} \tag{6–6}$$

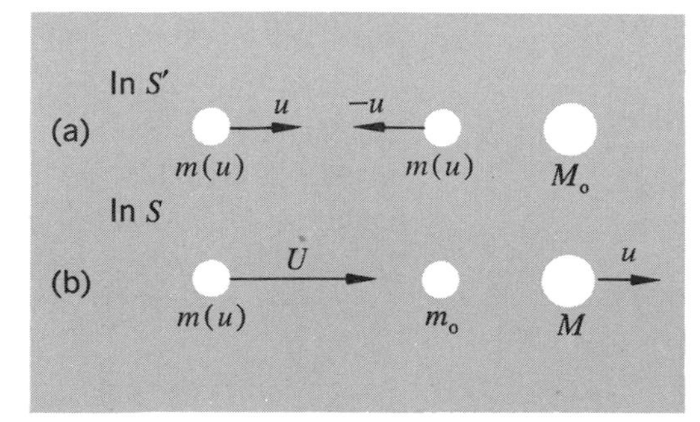

Fig. 6–2 Completely inelastic collision between two similar objects, observed (a) in the zero-momentum frame, (b) in a frame in which one of the objects is initially stationary.

Now let us write the statements of conservation of momentum and conservation of *mass*, from the standpoint of frame S. (As in the previous section, we suppose that the mass is some function of the speed):

Momentum:

$$m(U)U = Mu \tag{6–7}$$

Mass:

$$m(U) + m_0 = M \tag{6–8}$$

Eliminating M between these two equations we find

$$\frac{m(U)}{m_0} = \frac{u}{U - u} \tag{6–9}$$

Now Eq. (6–6) gives the connection between U and u; thus we can find the ratio $m(U)/m_0$ as an explicit function of U. From Eq. (6–6) we have

$$u^2 - 2(c^2/U)\,u + c^2 = 0$$

Therefore,

$$\begin{aligned} u &= \frac{c^2}{U} \pm \left[\left(\frac{c^2}{U}\right)^2 - c^2\right]^{1/2} \\ &= \frac{c^2}{U}[1 \pm (1 - U^2/c^2)^{1/2}] \end{aligned}$$

Since we must have $u \to U/2$ for $U \ll c$, we know that the negative sign should be chosen. (Appeal to the fact that the radical is approximately equal to $1 - U^2/2c^2$ for $U \ll c$.) Thus we have

$$u = \frac{c^2}{U}[1 - (1 - U^2/c^2)^{1/2}] \tag{6–10a}$$

Therefore,

$$U - u = \frac{c^2}{U}[U^2/c^2 - 1 + (1 - U^2/c^2)^{1/2}]$$

i.e.,

$$U - u = \frac{c^2}{U}(1 - U^2/c^2)^{1/2}[1 - (1 - U^2/c^2)^{1/2}] \tag{6–10b}$$

Substituting from equations (6–10) into (6–9), we have

$$\frac{m(U)}{m_0} = \frac{1}{(1 - U^2/c^2)^{1/2}} = \gamma(U) \tag{6–11}$$

which thus reproduces the form of the mass formula of Eq. (6–5).

This calculation involves more algebraic manipulation than the one in the previous section, but it is more satisfactory in several ways:

1. The collision considered is an extremely simple one, the motion being entirely along a single straight line.

2. The calculation is exact. One of the particles is, by definition, completely stationary in frame S before the collision (not just approximately so as it was in the previous case).

3. The explicit use of the mass-conservation equation leads naturally to the equivalence of mass and energy. We have, in essence, already developed the connection in these terms in Chapter 1, so the chief purpose of restating it here is to emphasize once again the intimate connection between the kinematics and the dynamics. From Eq. (6–11) we have

$$\begin{aligned} m(U) &= m_0(1 - U^2/c^2)^{-1/2} \\ &= m_0 + \tfrac{1}{2}m_0U^2/c^2 + \cdots \end{aligned}$$

Therefore,

$$m(U)c^2 = m_0c^2 + \tfrac{1}{2}m_0U^2 + \cdots \tag{6–12}$$

Noting that the second term on the right of Eq. (6–12) corresponds exactly to the classical kinetic energy of a particle of mass m_0 and speed U, we come to the familiar statement that the *total* energy of a particle of rest mass m_0 and speed U is given by

$$E = \frac{m_0c^2}{(1 - U^2/c^2)^{1/2}} = m(U)c^2 \tag{6–13}$$

with $E_0 = m_0c^2$ defining the rest energy of the particle.

4. By considering the collision further, we can demonstrate that the *consistent* use of a mass/velocity relation as given by Eq. (6–11) involves no contradictions. In Eq. (6–11) we have a statement of the mass of that colliding particle which, as observed in frame S, has speed U. Let us express this in terms of u, using Eq. (6–6); we have

$$\begin{aligned} 1 - U^2/c^2 &= 1 - \frac{4(u^2/c^2)}{(1 + u^2/c^2)^2} \\ &= \frac{(1 - u^2/c^2)^2}{(1 + u^2/c^2)^2} \end{aligned}$$

Therefore,

$$m(U) = \frac{(1 + u^2/c^2)}{(1 - u^2/c^2)}m_0$$

Substituting this value in Eq. (6–8), we find that the mass of the composite particle, as measured in frame S, is given by

$$M = \frac{2m_0}{1 - u^2/c^2} \qquad (6\text{–}14)$$

But in frame S this composite particle has speed u. On the basis of Eq. (6–11) we would infer that its rest mass should be M_0, where

$$M_0 = M(1 - u^2/c^2)^{1/2}$$

Using Eq. (6–14), this would give us

$$M_0 = \frac{2m_0}{(1 - u^2/c^2)^{1/2}} \qquad (6\text{–}15)$$

But now, consider the collision as described in the frame S'. Here the composite particle is indeed at rest, having been formed from the collision of two particles, each of rest mass m_0 and speed u. We are assuming that all the mechanical energy brought in by the colliding particles is retained within the composite particle. Thus we do not (and must not) assume that M_0 is equal to $2m_0$. Using the statement of conservation of mass as applied in frame S', we have

$$M_0 = 2m(u) \qquad (6\text{–}16)$$

which is identical with Eq. (6–15) if we accept the velocity dependence of mass as given in Eq. (6–11) and thus put $m(u) = \gamma(u)m_0$.

FURTHER REMARKS ON THE CONSERVATION LAWS

It should be very clear from all the preceding discussion that the momentum and energy conservation laws are not sacred; there is nothing, however, in our experience so far that has required their abandonment. It has been a pretty near thing at times—as, for instance, when the existence of the neutrino, a hitherto unobserved particle, was postulated by W. Pauli in 1930 to avoid giving up conservation of energy in beta decay. It took over 20 years before the neutrino was detected—but it was,[1] and our confidence in the conservation laws was still further strengthened thereby.

Clearly the conservation laws are not going to tell us the

[1]F. Reines, et al., *Phys. Rev.*, **92,** 830 (1953).

whole story by a long way. In the collision of two atomic particles, for example, we shall need detailed information about interatomic forces before being able to answer many of the important questions: Will the particles stick together? Or will they undergo an elastic collision? Or a partially inelastic collision? What will be the probability of scattering in a particular direction? What is the total effective target area that they present to each other? Will radiation be emitted after impact, and if so, of what kind? To provide the answers to these questions over a wide range of conditions is likely to require many man-years of research. Nevertheless, the basic conservation laws are the essential foundation for all else, and their generality makes them a powerful tool. What we shall do, then, is to hitch ourselves to the following statements (for an isolated system) and see where they lead us:

$$E_{\text{total}} = c^2 m_{\text{total}} = \text{constant} \tag{6–17}$$

$$\mathbf{p}_{\text{total}} = \text{constant} \tag{6–18}$$

We shall also be making use of the dynamic relations, already introduced in Chapter 1, that are derivable from equations (6–1):

$$E^2 = (cp)^2 + E_0{}^2 \qquad (\text{with } E = mc^2,\ E_0 = m_0c^2) \tag{6–19}$$

$$v = \frac{p}{m} = \frac{c^2 p}{E}$$

or (6–20)

$$\beta = \frac{v}{c} = \frac{p}{mc} = \frac{cp}{E}$$

Forces and accelerations will not enter this part of our discussion at all.

We shall begin with some collision problems involving photons, because their lack of any rest mass makes the equations less complicated.

ABSORPTION AND EMISSION OF PHOTONS

Absorption

Suppose that a stationary particle (e.g., an atom or nucleus) of mass (rest mass) M_0 is struck by a photon (quantum) of energy Q which is completely absorbed. The combined system will have mass M' and will recoil with a velocity v. Then we have

Conservation of energy:

$$E = M_0c^2 + Q = M'c^2$$

Conservation of linear momentum:

$$p = Q/c = M'v$$

Therefore,

$$M' = M_0 + Q/c^2$$
$$\beta = v/c = Q/(M_0c^2 + Q) \tag{6-21}$$

Note that, for $Q \ll M_0c^2$, we have simply $\beta \approx Q/M_0c^2$ —corresponding to a Newtonian type of calculation in which a body of invariable mass M_0 is given an impulse of magnitude Q/c by the photon.

Emission

Consider a stationary atom of mass M_0 that emits a photon of energy Q. This is already more complicated than the previous example, because the emitting atom undergoes a recoil. Let the recoiling atom have mass M' (and rest mass M_0') and velocity v. Then

$$E = M_0c^2 = M'c^2 + Q = E' + Q$$
$$p = 0 = M'v - Q/c = p' - Q/c$$

i.e.,

$$E' = M_0c^2 - Q$$
$$cp' = Q$$

We will solve these equations for Q by taking advantage of the relation between E' and p' for the recoiling atom. Using Eq. (6–19), we have

$$(M_0'c^2)^2 = (E')^2 - (cp')^2$$
$$= (M_0c^2 - Q)^2 - (Q)^2$$

i.e.,

$$(M_0'c^2)^2 = (M_0c^2)^2 - 2M_0c^2Q \tag{6-22}$$

Now M_0c^2 and $M_0'c^2$, the rest energies of the atom in its initial and final states, have certain definite values, and the difference between them is a well-defined, fixed energy. Let us therefore put

$$M_0'c^2 = M_0c^2 - Q_0 \tag{6-23}$$

Therefore,

$$(M_0'c^2)^2 = (M_0c^2)^2 - 2M_0c^2Q_0 + Q_0^2 \tag{6–24}$$

Combining Eqs. (6–22) and (6–24), we get

$$Q = Q_0\left(1 - \frac{Q_0}{2M_0c^2}\right) \tag{6–25}$$

Since the photon energy is proportional to the frequency, the corresponding frequency is lowered and the wavelength increased. Only if the emitting atom could somehow be prevented from recoiling would the total energy release Q_0 be conferred on the photon.

These results have important physical implications, because they place restrictions on the ability of atoms and nuclei to reabsorb their own characteristic radiations. Any element when suitably stimulated (as in an electric-discharge tube) emits a characteristic line spectrum—for example, the Balmer series of hydrogen. These lines are very sharp; that is to say, each line represents an extremely small spread of wavelengths about some average. This sharpness is an expression of the fact that the emitting atoms themselves cannot exist in states with any arbitrary energy but are limited to a series of sharp energy levels. The emission of a photon corresponds to a certain decrease of energy (or mass) of an atom, as described by Eq. (6–23), when the atom falls from a state A to a state B. The photon, however, is cheated out of a small fraction of this energy by the atomic recoil. Thus, if such a photon encounters another similar atom which is in its lower state B and at rest, there is not enough energy to raise the atom back to state A (and the situation is exacerbated by the fact that the absorption process in turn involves a recoil). If atomic energy states were *perfectly* sharp, and if emitting and absorbing atoms were both initially stationary, a vapor would thus be transparent to its own radiation. Of course, the situation we have described is unrealistic on two counts. Atomic energy-levels are not perfectly sharp, and the atoms of a vapor have thermal motions that can, if the velocities are right, nullify the effects of recoil. It turns out, in fact, that the thermal motions completely mask the effect in the case of visible light. But with the much more energetic photons that are ejected from nuclei as γ rays the recoil effect is relatively much greater [note that ac-

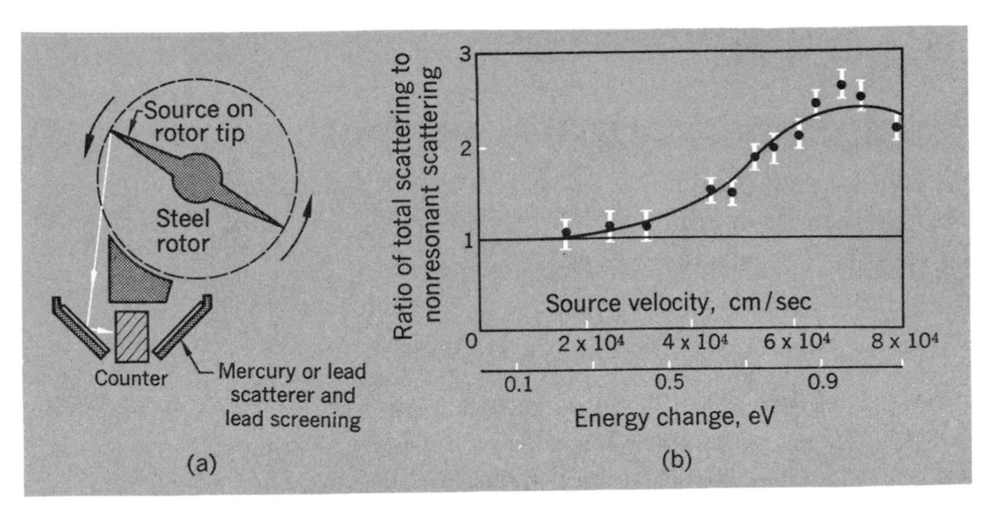

Fig. 6–3 Demonstration of nuclear resonant scattering by freely recoiling atoms of ^{198}*Hg.* (*a*) *Schematic diagram of apparatus.* (*b*) *Ratio of scattering by mercury to scattering by lead, plotted as a function of rotor tip speed* [*P. B. Moon, Proc. Phys. Soc.* (*London*) **A64,** *76* (*1951*)*; W. G. Davey and P. B. Moon, Proc. Phys. Soc.* (*London*), **A66,** *956* (*1953*)]. *The complete resonance curve could not be traced out, because of bursting of the steel rotors through centrifugal stresses at higher speeds.*

cording to Eq. (6–25) the *fractional* shift of energy is equal to $Q_0/2M_0c^2$ and so increases in proportion to Q_0 for a given value of M_0], and we shall briefly describe an experiment that demonstrated it.

The experiment was done by P. B. Moon, using an ultra-high-speed rotor. Figure 6–3(a) is a diagram of the arrangement. A radioactive source of excited atoms of ^{198}Hg was mounted on the tip of the rotor. Gamma rays (having a photon energy of 412 keV) emitted from the moving source fell upon a stationary target of mercury, from which a fraction of them were scattered into a counter and thereby detected. The scattering, like the absorption, is greatest when the energy is just right to lift the struck nucleus into a higher state. Experiment showed [cf. Fig. 6–3(b)] that this condition was reached at a source velocity of about 700 m/sec for ^{198}Hg.

It is clear that the nuclear recoil represents an extremely small fraction of the liberated energy. It is determined by the ratio $Q_0/2M_0c^2$ that appears in Eq. (6–25), and for the case under discussion we have

$$
\begin{aligned}
Q_0 &= 4.12 \times 10^5 \text{ eV} \\
M_0 &= 198 \text{ amu} = 3.28 \times 10^{-25} \text{ kg} \\
M_0c^2 &= 2.95 \times 10^{-8} \text{ joule} = 1.84 \times 10^{11} \text{ eV} \\
Q_0/2M_0c^2 &\approx 10^{-6}
\end{aligned}
$$

One can see that for photons of visible light, with $Q_0 \approx 2$ eV, the effect is five orders of magnitude smaller yet.

Each photon is involved in two recoils—once when it is emitted, and again when it is absorbed or scattered. The factor $(1 - Q_0/2M_0c^2)$ must thus be applied twice, and the velocity given to the source must be sufficient to make up for the fraction Q_0/M_0c^2 of the original excitation energy that is made unavailable in this way. It is possible to calculate the requisite velocity by appealing to the kinematic theory of the Doppler effect (Chapter 5). Since the nuclear recoil is very small, the frequency ν, and hence the energy $h\nu$, of the radiation is raised by the fraction v/c very nearly, as for any source moving toward an observer with a speed v that is much less than c. We therefore demand the relationship

$$v/c \approx Q_0/M_0c^2$$

Putting in the precise values of Q_0, M_0, and c, we have

$$v/c = 2.24 \times 10^{-6}$$

i.e.,

$$v \approx 670 \text{ m/sec}$$

The peak of the observed scattering curve corresponds quite closely to this figure.

THE MÖSSBAUER EFFECT

It would seem that an atomic or nuclear recoil ought to be an inescapable feature of photon emission. This, however, is not so. There is a most remarkable effect (named after R. Mössbauer, who received the Nobel prize in 1961 for discovering it) by which the recoil may to all intents and purposes be suppressed completely for atoms in a crystalline lattice. What happens is that, in some circumstances, the recoil momentum accompanying emission and absorption of a nuclear γ ray is taken up by all the atoms in the crystal, rather than by a single nucleus. Even if the crystal is minute by macroscopic standards, it contains so many atoms (e.g., about 10^{10} in a crystal that is $1\mu^3$ in size) that the ratio $Q_0/2M_0c^2$ in Eq. (6–25) is rendered almost vanishingly small. With the iridium isotope ^{191}Ir, for example (with which Mössbauer carried out his original experiments) the energy release Q_0 is 129 keV, or about 2×10^{-14} joule. If we take M_0 to be the mass of 10^{10} atoms of iridium, i.e., about 3×10^{-15} kg, we have

$$\frac{Q_0}{2M_0c^2} \approx \frac{2 \times 10^{-14}}{2 \times 3 \times 10^{-15} \times 9 \times 10^{16}} \approx 3 \times 10^{-17}$$

Now we remarked earlier that the existence of an atomic or nuclear recoil does not by itself guarantee that emitted photons cannot be reabsorbed by similar atoms. We can meaningfully state the value of an energy Q_0 that characterizes the difference of rest energies of the atom in its normal and excited states, but this is only a most probable value. The excited atom has a finite chance of being formed from the lower state if the energy input is close to Q_0, rather than being precisely equal to it. This characteristic "width" of an atomic or nuclear state is a very individual thing. If the width happens to be extremely small, the absorption is very sharply tuned, as it were. It is for such cases, where the slightest displacement of energies would make reabsorption impossible, that the Mössbauer effect becomes a spectacular phenomenon. To take ^{191}Ir as the example once again, the 129-keV recoilfree γ rays have energies varying over about 5×10^{-6} eV. Thus we have

$$\frac{\Delta Q}{Q_0} = \frac{5 \times 10^{-6}}{10^5} \approx 5 \times 10^{-11}$$

We may note first that this ratio is several orders of magnitude smaller than the ratio $Q_0/2M_0c^2$ if for M_0 we take the mass of just one atom. Thus reabsorption of these gamma rays would be impossible if either the emitting nucleus or the absorbing nucleus were free to recoil. On the other hand, the value of $\Delta Q/Q_0$ is many orders of magnitude *greater* than $Q_0/2M_0c^2$ if for M_0 we take the mass of billions of atoms. Thus if the excited nuclei of ^{191}Ir that emit the gamma rays, and the normal nuclei of ^{191}Ir that absorb them, are both embedded in crystal lattices, the recoil effects are quite negligible, and the absorption can take place.

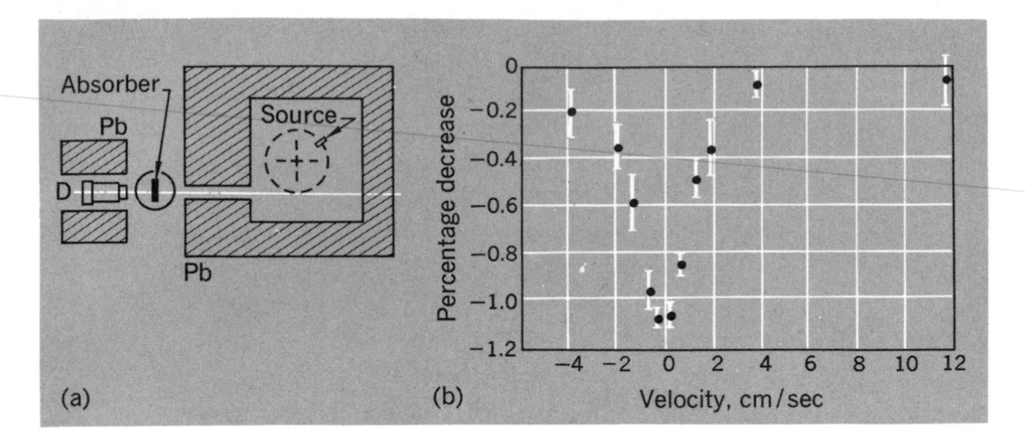

Fig. 6–4 Mössbauer effect. (a) Experimental arrangement. The detector D is shielded from the source except when the latter is moving in line with the long hole in the shield. (b) Percentage difference in intensity of 129-keV γ rays from ^{191}Ir as measured behind a resonance absorber (iridium) and a comparison absorber (platinum).

But now comes the really startling feature. The ratio $\Delta Q/Q_0$ is so tiny that by introducing just a small relative velocity between source and absorber it is possible, because of the resulting Doppler shift, to destroy the condition for reabsorption. In the case of ^{191}Ir we have

$$\frac{v}{c}\left(=\frac{\Delta Q}{Q_0}\right) \approx 5 \times 10^{-11}$$

Therefore,

$$v \approx 2 \text{ cm/sec}$$

This is, quite literally, little more than a snail's pace.

Figure 6–4(a) shows the experimental arrangement used by Mössbauer to demonstrate the effect. The source was mounted on a slowly rotating turntable. Figure 6–4(b) shows his results in this particular experiment. It is a truly remarkable phenomenon, which has since been observed for quite a number of emitters of low-energy γ rays. Some of them exhibit an even greater sensitivity than ^{191}Ir to the relative motion of source and absorber. But for details and for some of the fascinating applications, we must refer you elsewhere.[1] It is only because of

[1] See, for example, the excellent account by H. Lustig, *Am. J. Phys.*, **29**, 1–18 (1961).

the extraordinary interest of the Mössbauer effect that we have ventured to discuss it at all. Our original topic was the recoil accompanying photon emission, and this has been a whole section about the recoils that do not take place!

THE PHOTON ROCKET[1]

Enthusiasts for space travel have sometimes proposed the use of radiation as a propellant for a spaceship. The suggestion is motivated, at least in part, by the fact that the thrust of a rocket is proportional to the speed of the ejected fuel; thus there is, on the face of it, a lot to be gained by substituting c (3×10^8 m/sec) for the speeds of the order of 10^4 m/sec that represent the limit for chemical fuels. The suggestion might become very attractive if we could give a spacecraft a speed so near c that we could give travelers the benefit of a large time dilation on a lengthy interstellar trip.

Suppose we start with a rocket of rest mass m_0, which has a payload of rest mass equal to some fraction f of m_0. Starting from rest, we end up by having the payload traveling at speed v. A certain total amount of radiated energy, E_r, is traveling in the opposite direction. Using the energy and momentum conservation equations we thus have

$$E_{\text{total}} = m_0c^2 = \frac{fm_0c^2}{(1 - v^2/c^2)^{1/2}} + E_r$$

$$p_{\text{total}} = 0 = \frac{fm_0v}{(1 - v^2/c^2)^{1/2}} - \frac{E_r}{c}$$

Eliminating E_r between these equations, we have

$$m_0c^2 = \frac{fm_0c^2}{(1 - v^2/c^2)^{1/2}} + \frac{fm_0vc}{(1 - v^2/c^2)^{1/2}}$$

i.e.,

$$\gamma f + \gamma\beta f = 1 \qquad \text{where } \beta = v/c, \quad \gamma = (1 - v^2/c^2)^{-1/2} \qquad (6\text{–}26)$$

We want to design for a specified value of the time-dilation factor γ, so we shall solve Eq. (6–26) for f. Now β and γ are related by the equation

$$\gamma = (1 - \beta^2)^{-1/2}$$

[1]With acknowledgment to J. R. Pierce, *Proc. IRE*, **47**, 1053–1061 (1959).

Therefore,

$$\beta^2 = 1 - \frac{1}{\gamma^2}$$

$$\gamma\beta = (\gamma^2 - 1)^{1/2}$$

Thus Eq. (6–26) can be written

$$\gamma f + (\gamma^2 - 1)^{1/2} f = 1$$

which yields the following equation for f:

$$f^2 - 2\gamma f + 1 = 0$$

Suppose, for example, we put $\gamma = 10$, so that the time dilation makes a really significant difference in a long and tedious interstellar trip. Then we find $f \approx 0.05$. This looks encouraging. But if you are humanitarian (or a prospective space traveler yourself) you want to bring the travelers back. And that will mean going through three more stages of the same kind, because we must bring the spaceship to rest, reaccelerate it for the return journey, and finally stop it at the earth. The mass of the returning remnant can be only a fraction f^4 of the original mass m_0—i.e., in the case we have chosen, less than $10^{-5}m_0$. It isn't so very promising after all.

CREATION OF PARTICLES

Perhaps the most remarkable of all the possibilities suggested by the mass-energy equivalence is the creation of new particles if an adequate amount of energy is made available. A sizeable fraction of all activity in physics today is founded upon this phenomenon, especially for the purpose of creating and studying particles that are too short-lived to be found in nature. To create a particle of rest mass m_0 will clearly require an energy input of at least m_0c^2. In practice, however, more than this has to be used—for many cases a great deal more. There are two main reasons for this:

1. There appear to be fundamental conservation laws, over and above those for energy and momentum, which in many cases make it impossible to create only one new particle in a collision process. The most familiar of such laws is the conservation of electric charge. For example, the very first such

process to be discovered was the creation of an electron-positron pair from the energy of a γ-ray photon[1]:

$$\gamma \rightarrow e^- + e^+$$

Although, on energetic grounds alone, a γ ray of 0.51 MeV would suffice to provide the rest-mass energy of one electron, the only type of process that nature allows requires at least twice this amount.

Actually, although charge conservation applies invariably in these transmutations, it is by no means the only restriction. For example, one could envisage the creation of the constituents of a neutral hydrogen atom—one proton and one electron—using the energy of a single photon ($\geq$ 938 MeV). But this is not an observed process. It appears that many types of particles (including electrons, protons, and neutrons) cannot be created without calling into existence their so-called antiparticles—particles of the same rest mass, but with electric charge, magnetic moment, etc., of the opposite sign. The creation of a neutron, even though it is uncharged, does not occur without the simultaneous creation of an antineutron (differing from it in the sign of the magnetic moment).

2. The other reason that may step up the energy requirements for particle creation is a purely practical one. It arises from the fact that the creation process normally is made to take place by causing energetic collisions between preexisting particles. Thus, for example, positively charged π mesons (pions) can be made by bombarding a hydrogen target with high-energy protons:

$$P_1 + P_2 \rightarrow P + N + \pi^+$$

The colliding protons, P_1 and P_2, give rise to a proton, a neutron, and a pion, as indicated. (The π meson happens to be a particle that can be created singly, without an associated antiparticle.) Since a neutron and a proton have almost equal rest masses, the only new rest energy needed is that represented by the pion, about 140 MeV. But if the target proton P_2 is initially at rest and P_1 has a large momentum, a good deal of kinetic energy is locked up in motion of the system as a whole, and is unavailable for conversion into the rest mass of new particles.

It is clear that (2) is not a fundamental limitation in prin-

[1]First observed by C. D. Anderson in 1932. He was awarded the Nobel prize in 1936 for this research.

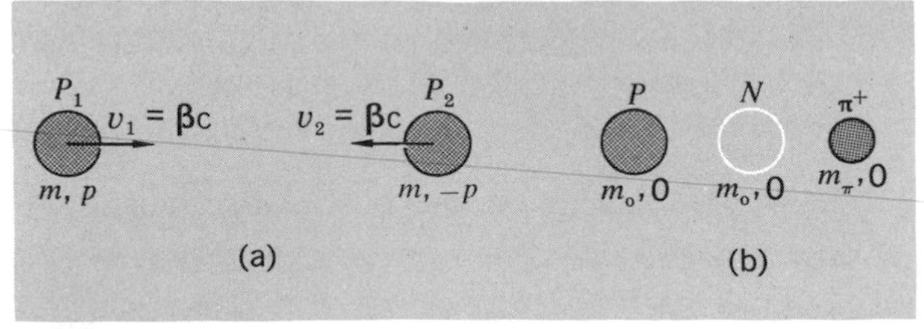

Fig. 6–5 (a) Two protons colliding with equal and opposite velocities in the zero-momentum frame. (b) The final state in this frame at the threshold for pion production, yielding a proton, a neutron, and a π^+ meson at rest.

ciple. If particles P_1 and P_2 could be made to collide with equal and opposite momenta, the amount of energy associated with the general motion of the system would be zero. All the kinetic energy of collision would then be available for particle creation. To produce colliding beams of particles traveling in opposite directions is technically a great deal harder than to have one beam striking a stationary target, but the payoff can be great, as we shall see.

Let us now consider in more detail some of these creation processes.

Pion production

Whether or not we have colliding beams, we can always *imagine* ourselves to be in a frame of reference where the total momentum is zero. Suppose we do this for two colliding protons, so that they have equal and opposite momenta, $\pm p$, and a total energy $2mc^2$ [Fig. 6–5(a)]. It is conceivable that in this zero-momentum frame we have a final state, as represented by Fig. 6–5(b), in which all particles are at rest. This will represent the most economical condition for particle creation, since nothing is wasted on kinetic energy, and will give us

$$E = 2mc^2 = 2m_0c^2 + m_\pi c^2$$

where m_0 is the rest mass of a nucleon—i.e., of either a proton or a neutron, disregarding the slight mass difference between them—and m_π is the rest mass of a charged pion.[1] Thus we have

$$\frac{m}{m_0} = 1 + \frac{m}{2m_0}$$

[1]Taking the electron mass as a unit, the proton mass is 1836.1 and the neutron mass is 1838.6, a difference of only 0.14%. The mass of a charged pion (+ or −) is $273.2m_e$. (Neutral pions are only $264.2m_e$.)

With $m_\pi = 273m_e$, $m_0 = 1837m_e$, this gives $m/m_0 = 1.074$, or $m_0/m = 0.93$. We can use this value of m_0/m to fix the speed (β) of each proton in the zero-momentum frame, for we have

$$m/m_0 = \gamma = (1 - \beta^2)^{-1/2} \tag{6–27}$$

$$\beta^2 = 1 - (m_0/m)^2 = 0.135$$

$$\beta \approx 0.37$$

Now if proton P_2 is actually at rest in the laboratory frame, the zero-momentum frame must have the speed β relative to the laboratory. Thus the proton P_1, which has the speed β in the zero-momentum frame, has a velocity β_1 in the laboratory frame given by

$$\beta_1 = \frac{\beta + \beta}{1 + \beta^2} = \frac{2\beta}{1 + \beta^2} \approx 0.65$$

according to the relativistic velocity-addition theorem [cf. Eq. (6–6)]. From this we have

$$\gamma_1 = (1 - \beta_1{}^2)^{-1/2} \approx 1.31$$

This means that the bombarding proton must have a *kinetic* energy of $(\gamma_1 - 1)m_0c^2$, or $0.31m_0c^2$. The rest energy of a nucleon is 938 MeV, so the kinetic energy required is about 290 MeV, or rather more than twice the rest energy of the created pion. It would have been precisely a factor 2 if we could have ignored the relativistic increase of mass with velocity for the protons. (Satisfy yourself that this is so.)

The bombarding energy as calculated here is what is called the *threshold energy* for the process. We know that anything less than this is insufficient, and in practice the bombardment is carried out at energies appreciably above threshold, because this enhances the efficiency of the process—i.e., the probability that in a proton-proton collision a pion will in fact be created. But this last statement raises questions beyond our present discussion, which is limited strictly to the collision dynamics and the calculation of threshold energies. Figure 6–6 is a bubble-chamber photograph showing the kind of evidence from which the occurrence of particle-creation events like these can be inferred.

Antiproton production

We mentioned earlier that the negatively charged antiproton cannot be produced without also calling into existence its own

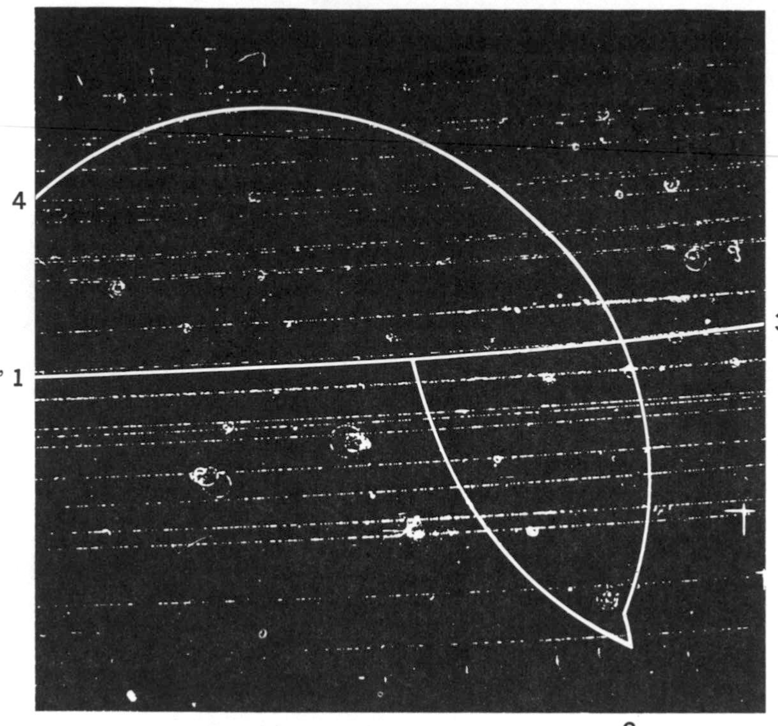

Fig. 6–6 Bubble-chamber photograph of the production of a π^+ meson through the process $p + p \rightarrow p + n + \pi^+$, using protons of about 3 GeV and a liquid-hydrogen bubble chamber. The incident proton (1) collides with a proton in the liquid hydrogen, producing a π^+ (2), a proton (3), and a neutron (which leaves no track). The π^+ is recognized by its subsequent decay giving a μ^+ meson (short track) which in turn decays, yielding a positron (4). All tracks are curved by a strong magnetic field applied perpendicular to the plane of the picture. Relevant tracks emphasized. (Brookhaven National Laboratory.)

antiparticle, i.e., an extra proton of the normal sort. It is possible to create a proton-antiproton pair by means of a proton-proton collision. The reaction can be written

$$P_1 + P_2 \rightarrow P_1 + P_2 + \overline{P} + P_3$$

where $\overline{P}$ is the antiproton and P_3 is the additional ordinary proton that appears with it. Hence, before even considering the unavailable kinetic energy, we know that at least twice the rest energy of the desired particle must be supplied. The most efficient conceivable way of making antiprotons through proton-proton collisions is to have a final state just like the one assumed in the pion-production problem, i.e., a state in which all the particles are at rest in some reference frame. The fact that we have four particles after the collision (three protons and an antiproton) instead of three does not make the calculation any harder. Indeed, the fact that all the particles have the same rest mass m_0 makes it arithmetically simpler to handle.

Figure 6–5(a) can again be used to represent the initial state in the zero-momentum frame. The total energy in this frame must be equal to (or greater than) the rest energy of four nucleons.

Hence at threshold we have

$$E = 2mc^2 = 4m_0c^2$$

Equation (6–27) thus gives us

$$\gamma = (1 - \beta^2)^{-1/2} = m/m_0 = 2 \qquad \text{whence} \quad \beta^2 = \tfrac{3}{4}$$

If we again assume that proton P_2 is a target proton at rest in the laboratory frame, the required velocity of P_1 in the laboratory is once more given by

$$\beta_1 = \frac{2\beta}{1 + \beta^2}$$

This time, instead of evaluating β_1 itself, let us go directly to the calculation of the corresponding value of γ_1:

$$\frac{1}{\gamma_1^2} = 1 - \beta_1^2 = \frac{(1 + \beta^2)^2 - (2\beta)^2}{(1 + \beta^2)^2} = \frac{(1 - \beta^2)^2}{(1 + \beta^2)^2}$$

Therefore,

$$\gamma_1 = \frac{1 + \beta^2}{1 - \beta^2} = \frac{1 + 3/4}{1 - 3/4} = 7$$

The bombarding proton must therefore have a total energy of $7m_0c^2$, i.e., a kinetic energy of $6m_0c^2$. Since the rest energy of a proton is 0.938 GeV (BeV), this calls for an incident proton beam of at least 5.62 GeV. It was for this reason that the University of California Bevatron, the first artificial source of antiprotons (1955), was designed to deliver protons of 6-GeV kinetic energy.

Pair production by photons

Figure 6–7 shows a beautiful example of the production of an electron-positron pair by a high-energy photon. The almost symmetrical spiraling of the tracks of the oppositely charged particles in a magnetic field shows that, in this particular event, the electron and positron were produced with nearly equal energies. One is left in no doubt that the photon that created this pair came in from the bottom of the picture, although being uncharged it left no track. But the process is more subtle than it looks. One might be tempted to imagine that a photon, all by itself, spontaneously transformed itself into an electron and a positron. But this is impossible on dynamical grounds alone, because linear momentum could not be conserved. To see that

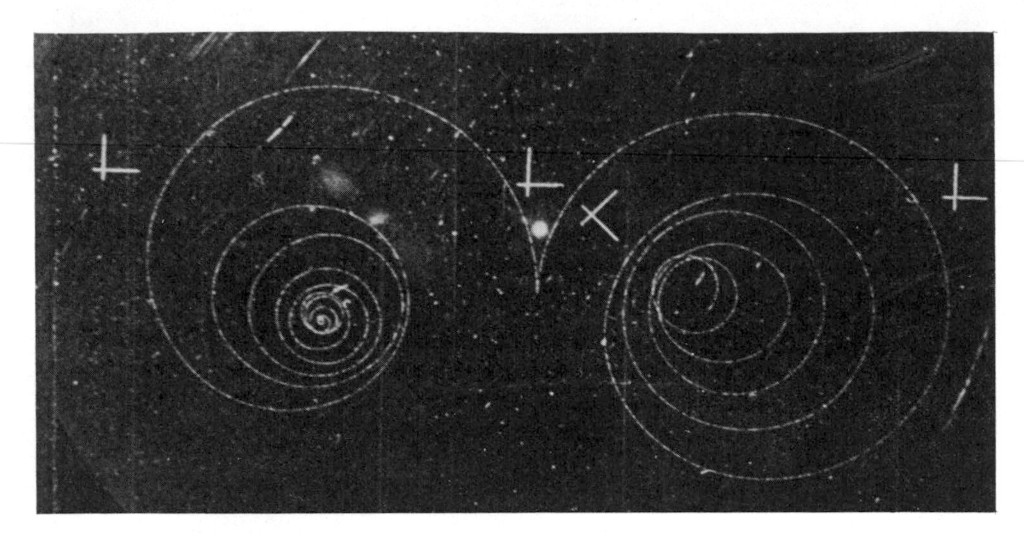

Fig. 6–7 Production of an electron-positron pair in a liquid hydrogen bubble chamber in a magnetic field. (Photo courtesy of Prof. Irwin Pless and the Harvard-M.I.T. bubble-chamber group.)

this is the case, consider the fact that whatever the velocities of the electron and positron may be in the laboratory, one can always find a reference frame in which their momenta are equal and opposite, so that the total linear momentum is zero. But one can never transform away the velocity c of a photon, and although the momentum of a photon will have different values in different frames, it can never be zero. Thus there must be another partner to the act of pair creation. This partner can be an electron or (more usually) an atomic nucleus.

If we imagine that the fourth particle is a nucleus, the analysis of the dynamics, at least in approximate terms, becomes quite simple. For a nucleus, being very massive, can take up a large amount of linear momentum without at the same time siphoning off a lot of kinetic energy—consider the Newtonian result that the kinetic energy of a particle of mass m and some given linear momentum p is equal to $p^2/2m$ and so can be made very small if m is very large. Thus it may be roughly true to say that all the energy of the photon goes into the electron and positron and that the nucleus takes care of the momentum balance.

In the photograph of Fig. 6–7, the pair production has occurred in a liquid-hydrogen bubble chamber. From the absence of any other visible recoil track, one can infer that the fourth partner in the process was a proton in one of the hydrogen atoms, rather than an electron. The rapid curling up of the electron and positron tracks shows that the particles lose energy at a rapid rate in traveling through the liquid hydrogen.

This kind of pair production by energetic photons is not limited to electrons, although the minimum photon energy for anything else is enormously greater than the 1.02 MeV that represents the threshold for electron/positron creation.

It is appropriate to mention here another process, which can be regarded as the inverse of pair creation. This is the mutual annihilation of a particle and its antiparticle with the production of radiant energy. A positron and an electron come together, and from the encounter two photons emerge. Why two and not one? Again we can recognize the requirements of momentum conservation at work. There is a reference frame (usually the laboratory itself) in which the total momentum of the electron-positron system is zero. At the very least we must have two photons (of equal energy, and traveling in opposite directions in this frame) to maintain the balance of both momentum and energy —at any rate if the electron and positron, when they annihilate, are effectively isolated from any other particles.

SCATTERING

As our next topic in this brief survey of relativistic collision problems, we shall consider two problems in elastic scattering—i.e., elastic collisions between particles. Since the study of scattering is one of the chief tools for investigating structures and forces on the atomic scale, a familiarity with the basic dynamics is important.

Elastic scattering of identical particles

As a first example we shall take the collision between two identical particles, e.g., two protons or two electrons. It is a well-known result of Newtonian mechanics that, if a stationary particle is struck elastically by an identical particle, the angle between their subsequent trajectories is exactly 90°. Figure 6–8 shows a pretty example of this for a proton-proton collision at low energy ($K \approx 5$ MeV, $v/c \approx 0.1$) that occurred when an incident proton struck a stationary proton in one of the hydrogen atoms in a photographic emulsion. Because of the increase of inertial mass with velocity, this result ceases to be true in the so-called *relativistic region* of energies; there is a squeezing forward of the directions so that the included angle becomes less than 90°.

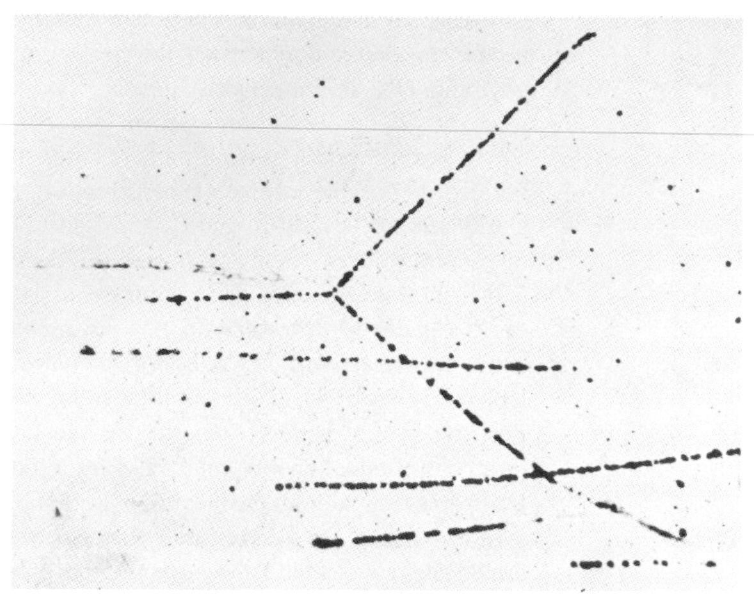

Fig. 6–8 Elastic scattering of an incident proton of about 5 MeV by an initially stationary proton in a photographic emulsion. The collision is "nonrelativistic" ($K/m_0c^2 \ll 1$) with a 90° angle between the tracks of the protons after collision. (From C. F. Powell and G. P. S. Occhialini, Nuclear Physics in Photographs, Oxford Univ. Press, New York.)

For simplicity we shall limit ourselves to considering the special case in which, after collision, the two particles (as observed in the laboratory frame S) travel symmetrically at equal angles to the direction of the incident particle. Let the incident particle have total energy E_1 and momentum $\mathbf{p}_1$, and let the momenta of the particles after collision be of magnitude p_2 at angles $\pm\theta/2$ to $\mathbf{p}_1$, as shown in Figure 6–9. Then by conservation of energy and momentum we have

$$E_1 + E_0 = 2E_2 \tag{6–28}$$

$$p_1 = 2p_2 \cos\frac{\theta}{2} \tag{6–29}$$

Also we have

$$c^2p_1{}^2 = E_1{}^2 - E_0{}^2 \qquad c^2p_2{}^2 = E_2{}^2 - E_0{}^2 \tag{6–30}$$

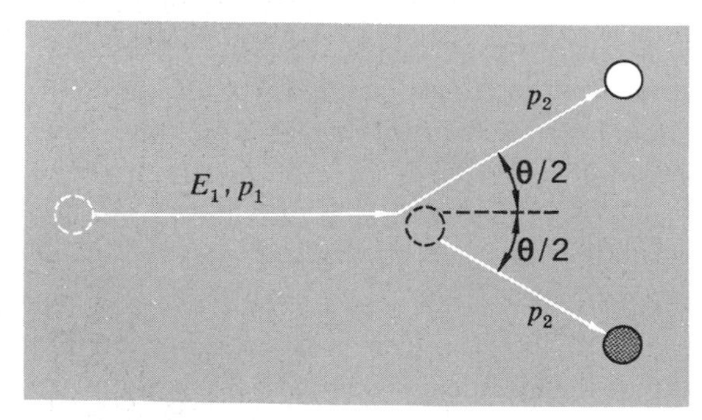

Fig. 6–9 Relativistic elastic collision of a particle with a similar particle initially at rest. The final state is assumed to be a symmetrical one in which the particles have equal speeds and hence make equal angles with the initial direction of particle 1.

It proves convenient to introduce the kinetic energy K_1 of the incident particle, so that we put

$$E_1 = E_0 + K_1$$

Using Eqs. (6–28) and (6–30) we then find

$$c^2p_1^2 = (E_0 + K_1)^2 - E_0^2 = K_1(2E_0 + K_1)$$

$$c^2p_2^2 = (E_0 + K_1/2)^2 - E_0^2 = K_1(E_0 + K_1/4)$$

Substituting these in Eq. (6–29) gives us

$$\cos^2\frac{\theta}{2} = \frac{2E_0 + K_1}{4E_0 + K_1}$$

Putting

$$\cos\theta = 2\cos^2\frac{\theta}{2} - 1$$

we find

$$\cos\theta = \frac{K_1}{4E_0 + K_1} \tag{6–31}$$

The change in the appearance of the collision as we go from low to high energies is nicely displayed in Eq. (6–31). For $K_1 \ll E_0$ we have $\cos\theta \to 0$, $\theta \to \pi/2$. For $K_1 \gg E_0$, we have $\cos\theta \to 1$, $\theta \to 0$. This relativistic compression of the scattering angles was first experimentally verified by F. C. Champion in 1932 for fast electrons (β particles).[1] Using a cloud chamber, he studied the elastic collisions of these electrons with the electrons of the atoms of the air in the chamber. Since that time the effect has become a commonplace in high-energy particle physics. Figure 6–10 shows a bubble-chamber photograph of a proton-proton

[1] F. C. Champion, *Proc. Roy. Soc.* (London), **A 136,** 630 (1932).

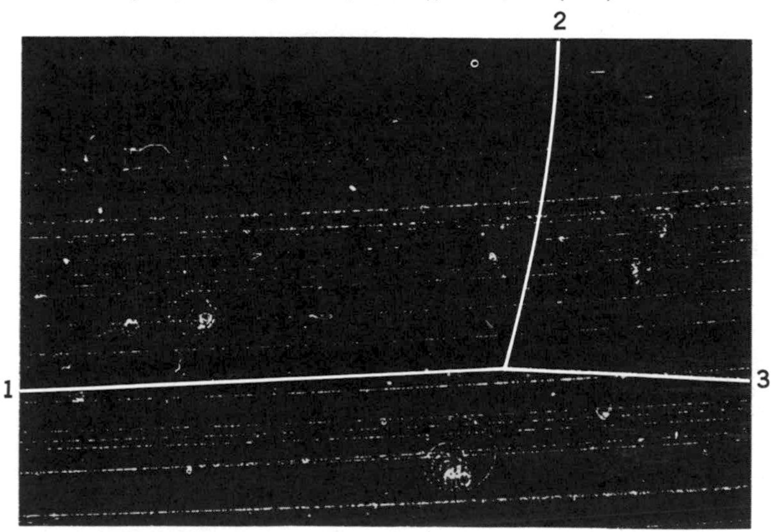

Fig. 6–10 Elastic proton-proton collision in a liquid-hydrogen bubble chamber, using incident protons of about 3 Gev. The incident proton enters at 1, and the two recoiling protons leave at 2 and 3. One cannot tell which of the latter was the incident proton. Relevant tracks emphasized. (*Brookhaven National Laboratory.*)

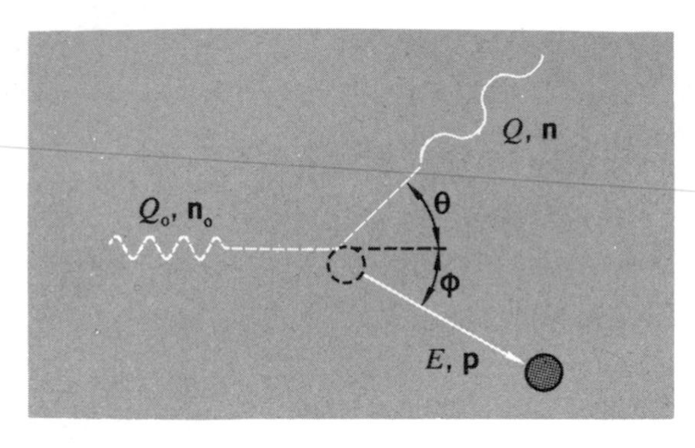

Fig. 6–11 Compton effect. An incident photon is scattered and degraded in energy as the result of an elastic collision with an initially stationary electron.

collision of this type, at an incident proton energy equivalent to several proton rest-masses.

The Compton effect

Of all the phenomena pointing to the corpuscular properties of photons, the Compton effect is perhaps the most direct and convincing. It is the collision of a photon with a free electron—which in practice means an electron loosely bound to an atom, so that it is effectively free. The collision is elastic, in the sense that no energy is siphoned off from kinetic energy into other forms, but because the electron recoils, the scattered photon has a lower energy, and hence a longer wavelength, than the incident photon. The systematic study of this phenomenon during the years 1919–1923 by A. H. Compton,[1] using X-ray photons, brought him a Nobel prize in 1927.

The Compton scattering process is an essentially relativistic collision, and can be described as follows. A photon of energy Q_0 strikes a stationary electron, which recoils in the direction φ (Fig. 6–11). The photon is scattered in the direction θ with energy Q. Conservation of energy and momentum give us the following:

$$Q_0 + m_0c^2 = E + Q \tag{6–32}$$

$$\mathbf{n}_0 Q_0/c = \mathbf{n}Q/c + \mathbf{p} \tag{6–33}$$

where E and $\mathbf{p}$ are the energy and momentum of the recoiling electron. If we are interested in the scattered photon and not in the electron, we can proceed as follows:

$$(Q_0 - Q) + m_0c^2 = E$$

$$(\mathbf{n}_0 Q_0 - \mathbf{n}Q) = c\mathbf{p}$$

[1] A. H. Compton, *Phys. Rev.*, **22,** 409 (1923).

where $\mathbf{n}_0$ and $\mathbf{n}$ are unit vectors in the initial and final photon directions, as shown. Square each of the above (i.e., form the scalar product of each side with itself in the second case):

$$(Q_0 - Q)^2 + 2(Q_0 - Q)m_0c^2 + (m_0c^2)^2 = E^2 \tag{6–34}$$

$$Q_0{}^2 - 2Q_0Q\cos\theta + Q^2 = c^2p^2 \tag{6–35}$$

Subtracting Eq. (6–34) from (6–35),

$$2Q_0Q(1 - \cos\theta) - 2(Q_0 - Q)m_0c^2 = 0$$

Therefore,

$$\frac{1}{Q} - \frac{1}{Q_0} = \frac{1}{m_0c^2}(1 - \cos\theta)$$

If the quantum energy is Q, the wavelength is given by

$$Q = h\nu = \frac{hc}{\lambda}$$

Thus in terms of wavelength the Compton effect is described by the following equation:

$$\lambda - \lambda_0 = \frac{h}{m_0c}(1 - \cos\theta) \tag{6–36}$$

For electrons, $h/m_0c = 0.02426$ Å, or 2.4×10^{-10} m. What Compton did was to establish that the scattered X-ray wavelength conformed to Eq. (6–36), both in its angular dependence and in the absolute size of the shift.[1] Figure 6–12 is a graph constructed from Compton's published data. It remained a matter of great interest, however, to demonstrate the ballistic nature of the collision by showing that the recoiling electron appeared simultaneously with the photon, and in a direction φ uniquely defined by the dynamics. The latter feature was convincingly demonstrated by Cross and Ramsey in 1950, using incident photons (γ rays, in this instance) with a sharply defined energy of 2.6 MeV. The experiment confirmed that the angle between photon and electron after scattering had the theoretical value within narrow limits (see Fig. 6–13). The coincidence in time between the particles in a Compton scattering process has

[1]The latter point is important, because even on a classical wave picture of radiation one can picture a free electron as being given a velocity under the action of radiation pressure. Radiation scattered from it would then be Doppler-shifted with the same angular variation as that given by Eq. (6–36). But the size of the shift would not be sharply defined, because the electron velocity would increase continuously from zero.

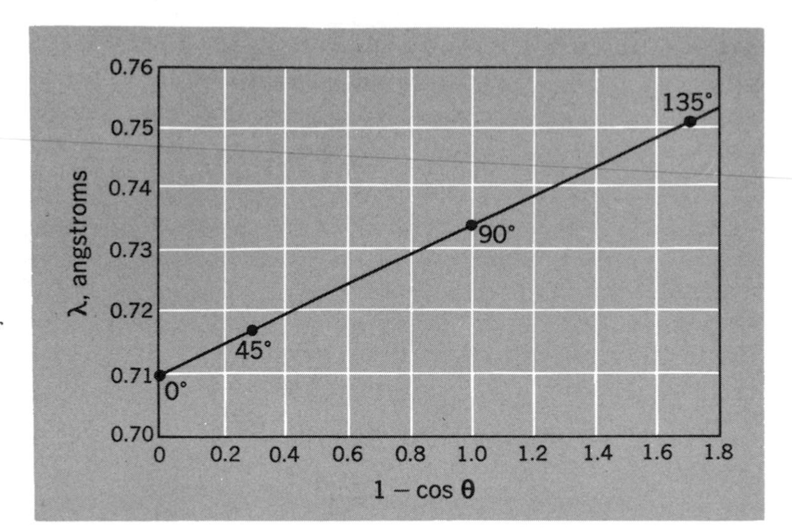

Fig. 6–12 Result of A. H. Compton's original experiment on the scattering of X rays by effectively free electrons.

been established as being good to 10^{-11} sec or less.[1] Although 10^{-11} sec is not particularly short on the time scale of atomic processes, the close association of an electron with each scattered photon is clearly shown by this result.

[1] Z. Bay, V. P. Henri, and F. McLernon, *Phys. Rev.*, **97,** 1710 (1955).

Fig. 6–13 Schematic diagram of the experiment by Cross and Ramsey to verify the complete dynamics of a Compton scattering process. A narrow beam of 2.6-MeV gamma rays, collimated by lead blocks, fell on a thin foil of beryllium. Photons scattered at 30° were detected by the flashes of light they produced in an anthracene crystal. The associated electrons recoiling at 31.3° were similarly detected, and a coincidence circuit recorded these events if they were simultaneous to within 1.5 × 10^{-8} sec. Each count then represented a single collision between a photon and an electron. [After W. G. Cross and N. F. Ramsey, Phys. Rev., **80,** *929 (1950).]*

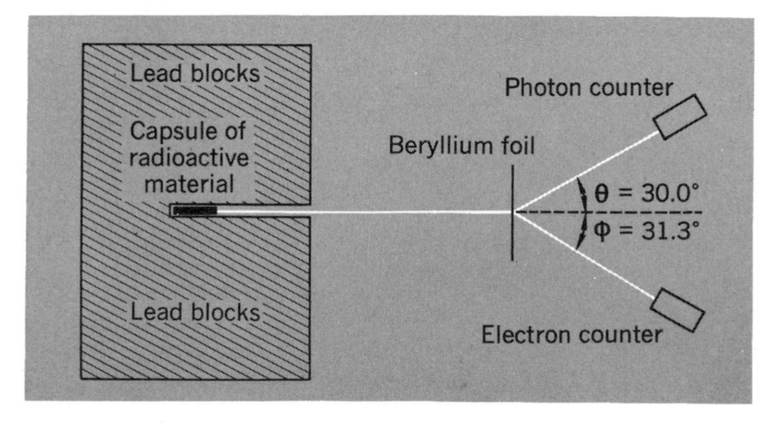

In Chapter 5 we treated the Doppler effect as a matter of pure kinematics. Our brief use of the effect in discussing γ-ray scattering earlier in this chapter was also essentially kinematic, although we inferred a dynamical consequence by appealing to the proportionality of energy to frequency for photons. But now we shall apply the energy-momentum arguments to the general problem of photon emission from a moving particle.

Suppose that a particle of mass M and momentum $\mathbf{p}$ emits a photon of energy Q' at an angle θ to the initial direction of motion of M, as shown in Fig. 6–14. The particle, after the emission, has a different mass M', and a momentum $\mathbf{p}'$ that may differ in both magnitude and direction from the initial momentum $\mathbf{p}$. Thus we have

$$Mc^2 = M'c^2 + Q' \tag{6–37}$$

$$\mathbf{p} = \mathbf{p}' + \mathbf{n}Q'/c \tag{6–38}$$

(where $\mathbf{n}$ is a vector of unit length along the direction θ). We shall fix attention on the photon, rather than on the recoiling particle, and we shall rearrange the equations to eliminate the unwanted quantities. Here is how to do it. Rewrite Eqs. (6–37) and (6–38) as follows:

$$Mc^2 - Q' = M'c^2 \tag{6–39}$$

$$c\mathbf{p} - \mathbf{n}Q' = c\mathbf{p}' \tag{6–40}$$

Square both sides of Eq. (6–39), and form the scalar product of each side of Eq. (6–40) with itself:

$$(Mc^2)^2 - 2Mc^2Q' + (Q')^2 = (M'c^2)^2 \tag{6–41}$$

$$(c\mathbf{p} - \mathbf{n}Q')\cdot(c\mathbf{p} - \mathbf{n}Q') = (c\mathbf{p}')\cdot(c\mathbf{p}')$$

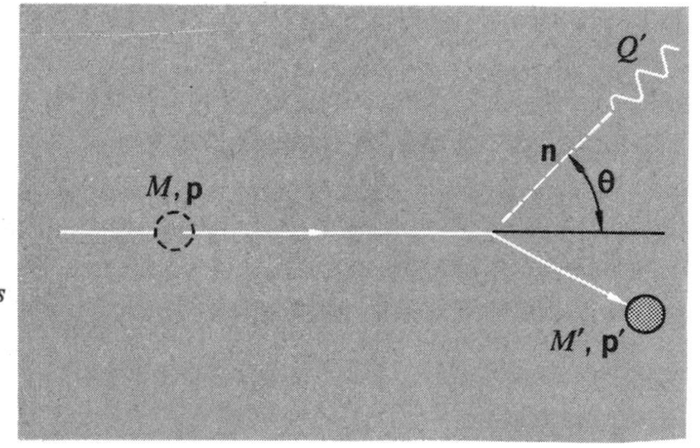

Fig. 6–14 Doppler effect, considered dynamically in terms of energy and momentum conservation.

i.e.,

$$(cp)^2 - 2cQ'(\mathbf{p} \cdot \mathbf{n}) + (Q')^2 = (cp')^2$$

or

$$(cp)^2 - 2cpQ' \cos \theta + (Q')^2 = (cp')^2 \tag{6–42}$$

Subtract Eq. (6–42) from Eq. (6–41):

$$[(Mc^2)^2 - (cp)^2] - 2Mc^2Q'[1 - (p/Mc) \cos \theta] = (M'c^2)^2 - (cp')^2$$

But

$$(Mc^2)^2 - (cp)^2 = (M_0c^2)^2$$
$$(M'c^2)^2 - (cp')^2 = (M_0'c^2)^2$$

and

$$p/Mc = v/c = \beta$$

Therefore,

$$(M_0c^2)^2 - 2Mc^2Q'(1 - \beta \cos \theta) = (M_0'c^2)^2 \tag{6–43}$$

As in our earlier, more restricted discussion of photon emission, we denote the decrease of rest energy by Q_0, so that

$$M_0'c^2 = M_0c^2 - Q_0$$

Therefore,

$$(M_0c^2)^2 - (M_0'c^2)^2 = 2M_0c^2Q_0 - Q_0^2$$

Hence Eq. (6–43) gives us

$$2Mc^2Q'(1 - \beta \cos \theta) = 2M_0c^2Q_0(1 - Q_0/2M_0c^2)$$

or

$$Q' = \frac{M_0}{M} \frac{Q_0(1 - Q_0/2M_0c^2)}{1 - \beta \cos \theta}$$

Now $M_0/M = (1 - \beta^2)^{1/2}$, so that we have, finally,

$$Q'(\beta, \theta) = Q \frac{(1 - \beta^2)^{1/2}}{1 - \beta \cos \theta} \tag{6–44}$$

where Q is the photon energy for $v = 0$, as given by Eq. (6–25) in our analysis of this special case.

In Eq. (6–44) we have a statement of the energy of a photon emitted in any direction θ from a moving source, with the effects of recoil included. It is identical in form with our purely kinematic result relating the received frequency ν' to the proper frequency ν for a moving source:

$$\nu' = \nu \frac{(1 - \beta^2)^{1/2}}{1 - \beta \cos \theta}$$

[see Eq. (5–17)].[1]

The above analysis may appear rather formidable at first sight, but it really isn't. The initial statements of energy and momentum conservation in Eqs. (6–37) and (6–38) are completely straightforward. And thereafter it is chiefly a matter of making systematic use of the extremely powerful relation

$$E^2 - (cp)^2 = E_0{}^2$$

by which the energy and momentum of a given particle can be combined into that constant quantity, the rest energy.

In optical spectroscopy the Doppler effect manifests itself as a small though quite measurable broadening of spectral lines, resulting from the random thermal motions of the emitting atoms. In high-energy particle physics, however, it will frequently happen that photons are emitted from unstable particles traveling with speeds comparable to c, leading to Doppler shifts that are large fractions of the total energy of the photon.

PROBLEMS

6–1 Suppose one photon has an energy of 200 MeV and is traveling along the x axis. Suppose another has an energy of 100 MeV and is traveling along the y axis. What is the total energy of this system?

[1]Should one simply say "Of course" upon seeing that the photon energy and the frequency depend in exactly the same way on the velocity and direction of the source? Certainly the quantum picture of electromagnetic radiation would be in a bad way if this agreement did not hold. And perhaps one can put it more strongly. Suppose you were a firm believer in relativity theory and then began speculating about the quantization of radiant energy. The comparison of the expressions for Q' and ν' would almost force you to conclude that quantum energy and frequency must be proportional to one another—in other words, if the quantum idea has any substance, then E/ν = constant is a necessity. (Relativity provides no hint, of course, as to what the value of the constant should be.) It would be tempting to imagine that Einstein himself followed some such line of thought when in the same year (1905) he published both the relativity theory and the first clear statement that $E/\nu = h$ for photons. (His writings do not, however, support any such assumption.)

The total momentum? If a single particle had this same total energy and momentum, what would be its mass? In what direction would it be traveling? With what speed?

6–2 A particle as observed in a certain reference frame has a total energy of 5 GeV and a momentum of 3 GeV$/c$ (i.e., cp, which has the dimension of energy, is equal to 3 GeV).

(a) What is its energy in a frame in which its momentum is equal to 4 GeV$/c$?

(b) What is its rest mass in amu?

(c) What is the relative velocity of the two reference frames?

6–3 A particle of rest mass m_0 and kinetic energy $2m_0c^2$ strikes and sticks to a stationary particle of rest mass $2m_0$. Find the rest mass M_0 of the composite particle.

6–4 (a) A photon of energy E collides with a stationary particle of rest mass m_0 and is absorbed. What is the velocity of the resulting composite particle?

(b) A particle of rest mass m_0 moving at a speed of $4c/5$ collides with a similar particle at rest and forms a composite particle. What is the rest mass of the composite particle and what is its speed?

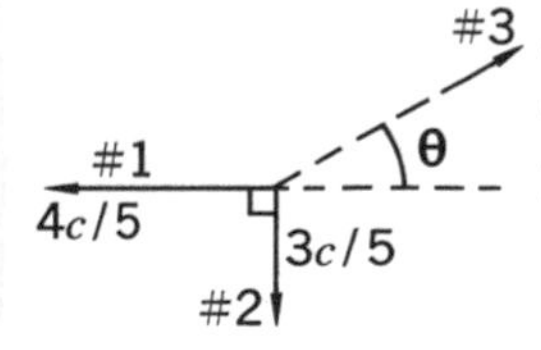

6–5 A particle of rest mass M_0 is at rest in the laboratory when it decays into three identical particles, each of rest mass m_0. Two of the particles (labeled #1 and #2) have velocities and directions as shown.

(a) Calculate the direction and the speed of particle #3.

(b) Find the ratio M_0/m_0.

6–6 (a) A fast-spinning wheel has a recoilless γ-ray source at its center and an appropriate absorber at its rim. Choose wheel speed and size so that the iridium activity used by Mössbauer would yield a just detectable transverse Doppler effect. What is the rim speed?

(b) If the *absorber* were at the center and the source were at the rim of the wheel, how would the analysis (and the effect itself) be changed?

(c) Suppose now that both source and detector are on the rim of the wheel, at opposite ends of a diameter. A photon leaving the source at position S_1 (see the figure) must travel along a chord so as to enter the absorber at a later time at position A_2. The photons make angles with the directions of motion of source and absorber which differ from 90° by α. The analysis thus requires the full Doppler formula, Eq. (5–17) or (6–44). What happens in this case?

(d) It is possible to consider all three situations from the standpoint of the frame of the rotating wheel. Both source and detector are then stationary in every case, but the photons move in an effective gravitational field of magnitude $\omega^2 r$ (and therefore increasing proportionally to r from the center of the wheel outward) associated with the centrifugal force. You can use the equivalence principle and consider the energy change that this field gives to the photons at any particular value of r.

6–7 A "photon rocket" uses pure radiation as the propellant. If the initial and final rest masses of the rocket are M_i and M_f, show that the final velocity v of the rocket relative to its initial rest frame is given by the equation

$$M_i/M_f = [(c + v)/(c - v)]^{1/2}$$

6–8 A thrust-beam space vehicle works by bearing a sort of sail which feels the push of a strong steady laser light beam directed at it from earth. If the sail is perfectly reflecting, calculate the mass of light required to accelerate a vehicle of rest mass M_0 up to a fixed value of gamma.

6–9 A laser with a mass of 10 kg is in free space with its beam directed toward the earth. The laser continuously emits 10^{20} photons/sec, of wavelength 6000 Å (as measured in its own rest frame). At $t = 0$ the laser is at rest with respect to the earth.

(a) Initially, how much radiant energy per second is received at the earth?

(b) The radiation emitted toward the earth causes the laser to recoil away from the earth. What is the velocity of the laser relative to the earth after 10 years (laser time) have elapsed?

(c) At the time when the laser is moving with velocity βc relative to the earth, how much less is the rate at which energy is received on earth than the original rate when $\beta = 0$? Evaluate this for $t = 10$ years (laser time).

(d) Show how an observer on earth can explain (to a first approximation) the continually decreasing rate of reception in terms of energy conservation.

6–10 An atom in an excited state of energy Q_0 above the ground state moves toward a scintillation counter with speed v. The atom decays to its ground state by emitting a photon of energy Q (as recorded by the counter), coming completely to rest as it does so. If the rest mass of the atom is m, show that $Q = Q_0[1 + (Q_0/2mc^2)]$.

6–11 The neutral π meson (π^0) decays into two γ rays (and nothing else). If a π^0 (whose rest mass is 135 MeV) is moving with a kinetic energy of 1 GeV:

(a) What are the energies of the γ rays if the decay process causes them to be emitted in opposite directions along the pion's original line of motion?

(b) What angle is formed between the two γ rays if they are emitted at equal angles to the direction of the pion's motion?

6–12 An antiproton $\tilde{p}$ of kinetic energy $\frac{2}{3}$ GeV strikes a proton p which is at rest in the laboratory. They annihilate (reaction: $\tilde{p} + p \rightarrow \gamma_1 + \gamma_2$), yielding two photons which emerge from the reaction traveling forward or backward on the line along which the antiproton entered. Take the rest energy of the proton and the antiproton to be 1 GeV each.

(a) What energies do the photons have?

(b) In which direction is each photon heading?

(c) As measured in a reference frame attached to the incoming $\tilde{p}$, what energy does each of the photons have?

6–13 A positron with kinetic energy 0.51 MeV collides inelastically with an electron at rest, forming a positronium atom which recoils freely. The electron and positron forming the positronium annihilate in flight, producing two γ rays.

(a) What is the speed of the positronium atom?

(b) What is the maximum possible energy of an annihilation photon so produced?

6–14 Show that the following processes are dynamically impossible:

(a) A single photon strikes a stationary electron and gives up all its energy to the electron.

(b) A single photon in empty space is transformed into an electron and a positron.

(c) A fast positron and a stationary electron annihilate, producing only one photon.

6–15 (a) If a proton of kinetic energy 437 MeV collides elastically with a proton at rest, and the two protons rebound with equal energies, what is the included angle between them? [R. B. Sutton et al., *Phys. Rev.*, **97,** 783 (1955), find 84.0° ± 0.2° for the experimental result.]

(b) If the incoming proton has a total energy of 33 GeV what is the included angle between them?

6–16 The usual theory of the Compton effect considers a stationary free electron being struck by a photon, resulting in a scattered photon of lower energy. Suppose that a photon (of energy Q) has a head-on collision with a *moving* electron (of rest mass m_0; see the figure). What initial velocity must the electron have if the collision results in a photon recoiling straight backward with the same energy Q as the incident photon?

6–17 A stream of very high energy photons ($\gg$ 10 MeV) is fired at a block of matter. Show that the energy Q of the photons scattered directly backward is essentially independent of the energy of the incident photons. What is the value of Q?

6–18 (a) A photon of energy $h\nu$ collides elastically with an electron at rest. After the collision the energy of the photon is $h\nu/2$, and it travels in a direction making an angle of 60° with its original direction. What is the value of ν? What sort of photon is this?

(b) A photon of energy $h\nu$ collides with an excited atom at rest. After the collision the photon still has energy $h\nu$, but its direction has changed by 180°. If the atom is in its ground state after the collision, what was its initial excitation energy?

6–19 A high-energy photon strikes and is scattered by a proton that is initially stationary and completely free to recoil. The proton is observed to recoil at an angle $\varphi = 30°$ with a kinetic energy of 100 MeV.

(a) What was the energy of the incident photon?

(b) What are the direction and energy of the scattered photon?

6–20 It is believed that Compton scattering by starlight quanta may be a mechanism for the energy degradation of high-energy electrons in interstellar space. An experiment has been proposed [R. Milburn, *Phys. Rev. Letters*, **10,** 75 (1963)] in which this phenomenon can be observed directly in the laboratory by scattering a high-energy electron beam against the intense flux of visible photons produced by a typical laser. Show that for such a process the laboratory energy of the scattered photon is given to an excellent approximation ($\beta \approx 1$) by

$$E_2 \approx \gamma mc^2 \left[\frac{\lambda(1 - \beta \cos \theta_0)}{1 + \lambda(1 - \cos \theta_0)}\right]$$

where $\lambda = 2\gamma E_1/mc^2$ (E_1 is the energy of the incident photon) and θ_0 is the photon scattering angle in the electron rest frame. Show also that θ, the photon scattering angle in the laboratory, is given by

$$\tan \theta = \frac{\sin \theta_0}{\gamma(\cos \theta_0 - \beta)}$$

If the incident electron beam is accelerated to an energy of 6.00 GeV, and the photons are generated by a 6943-Å ruby laser (so that their energy is 1.79 eV), what is the maximum energy of the scattered photons?

In relativity, we must complete the law of conservation of momentum by extending it to include the time component . . . The conservation of energy is the fourth equation which goes with the conservation of momentum to make a valid four-vector relationship in the geometry of space and time.

R. P. FEYNMAN, *Lectures on Physics* (1963)

7 More about relativistic dynamics

IN THIS CHAPTER we shall be discussing two main topics. The first of these is a more extended discussion of momentum and energy, with particular emphasis on the transformation of these quantities between two inertial frames. The second topic is the concept of force in relativistic dynamics—the way in which it is defined, its transformations, and the limitations on its usefulness. We begin with an important invariant that can be constructed from the measured values of momentum and energy in a given frame.

AN ENERGY-MOMENTUM INVARIANT AND ITS USE

The recipe for all that we have done so far is to assert that the energy and the linear momentum are separate constants for any self-contained system. In applying this, however, we take it for granted that we choose a particular inertial frame and stick to it throughout the calculation. What we shall do now is to ask ourselves how the measurements in *different* frames of E and p for a system of particles are related. The answer to this question is essentially contained in the basic statements of the relativistic momentum and energy of a single particle:

$$\mathbf{p} = \gamma m_0 \mathbf{v}$$
$$E = \gamma m_0 c^2$$

We have already seen how these can be related through the equation

$$E^2 = (cp)^2 + E_0{}^2$$

in which E_0 $(= m_0c^2)$ is the rest energy of the particle.

Now a simple way of restating this result is that, if a particle has rest energy E_0 (i.e., total energy E_0 as measured in the frame in which its momentum is zero), then its energy and momentum as measured in any other frame can be combined to form an invariant quantity as follows:

$$E^2 - (cp)^2 = E_0{}^2$$

(a relation that we exploited on a number of occasions in Chapter 6). Since this holds for E and p as measured in *any* frame, the measures of energy and momentum for a particle in *any two* frames are related according to the equation

$$E^2 - (cp)^2 = (E')^2 - (cp')^2 = E_0{}^2 \qquad (7\text{–}1)$$

where $E_0{}^2$ thus plays the role of an invariant dynamic property of the particle.

It turns out that Eq. (7–1) applies not merely to a single particle, but to any arbitrary collection of particles, in the following way. If, as measured in any given frame of reference, the sum of the energies of the particles is E and the vector sum of all their momenta is of magnitude p, then the value of $E^2 - (cp)^2$ has the same value as the corresponding combination $(E')^2 - (cp')^2$ as measured in any other frame. This invariant value is equal to the square of the total energy E_0 of all the particles as measured in a frame in which the vector sum of the momenta is zero.

Note especially that, in this extended form of Eq. (7–1), the energy E_0 is not, in general, merely a sum of rest energies. The collection of particles considered may have all kinds of motions relative to one another; there need not exist any frame in which they are all at rest.

A formal justification of this application of Eq. (7–1) to an arbitrary group of particles will be developed in the next section.

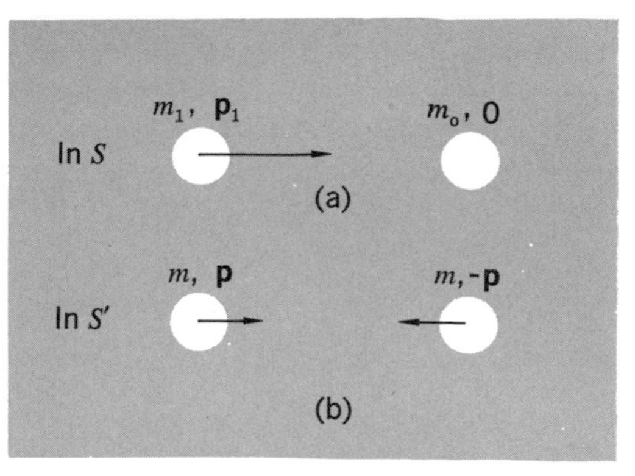

Fig. 7–1 Proton-proton collision (a) in a laboratory frame in which one proton is initially stationary; (b) in the zero-momentum frame.

We can, however, do something toward making it plausible right away by pointing out that we frequently, and justifiably, treat a collection of particles as though it were a single particle. Consider, for example, an argon atom containing numerous electrons in states of rapid motion, and having at its center a nucleus, itself a composite of neutrons and protons with large kinetic energies. We have no hesitation in describing this atom, from the standpoint of the kinetic theory of gases, as a single particle endowed with a certain velocity. And the theorem of the inertia of energy makes it all the easier to think of this complicated structure as being describable in terms of a single mass possessed of a certain momentum, despite our awareness of its internal structure.

A good example of the usefulness of Eq. (7–1) in its generalized form is provided by the antiproton creation problem that we have already treated by a different method. Figure 7–1(a) shows the initial proton-proton system as viewed in the laboratory; Figure 7–1(b) shows the same system in its own zero-momentum frame. In this latter frame the total energy $2mc^2$ of the colliding protons must be at least enough to represent four proton rest masses. Under these minimal conditions the final state, as observed in S', will consist of three protons and one antiproton, all with zero kinetic energy. Hence we have the following relationships:

In S':

$$E' = 2mc^2 = 4m_0c^2 \qquad p' = 0$$

In S:

$$E = (m_1 + m_0)c^2 \qquad p = p_1$$

Thus, by Eq. (7–1),

$$(m_1c^2 + m_0c^2)^2 - (cp_1)^2 = (4m_0c^2)^2$$

Therefore,

$$(m_1c^2)^2 + 2(m_1c^2)(m_0c^2) + (m_0c^2)^2 - (cp_1)^2 = 16(m_0c^2)^2$$

But

$$(m_1c^2)^2 - (cp_1)^2 = (m_0c^2)^2$$

since this represents Eq. (7–1) as applied to a single proton. Hence

$$2(m_1c^2)(m_0c^2) + 2(m_0c^2)^2 = 16(m_0c^2)^2$$

or

$$m_1c^2 = 7m_0c^2$$

We are thus enabled to calculate the necessary energy for the reaction without considering any of the velocities involved, in contrast to the treatment that we presented in Chapter 6 for this same problem.

LORENTZ TRANSFORMATIONS FOR ENERGY AND MOMENTUM

We shall consider the energy and momentum of a particle as measured with respect to two reference frames that are related via the usual Lorentz transformations. Let its velocity be **u** as measured in S, and **u′** as measured in S' (Fig. 7–2). Then by the velocity-transformation formulas we have

$$\left.\begin{aligned} u_x' &= \frac{u_x - v}{1 - vu_x/c^2} \\ u_y' &= \frac{u_y/\gamma(v)}{1 - vu_x/c^2} \end{aligned}\right\} \quad \text{where } \gamma(v) = (1 - v^2/c^2)^{-1/2} \qquad (7\text{–}2)$$

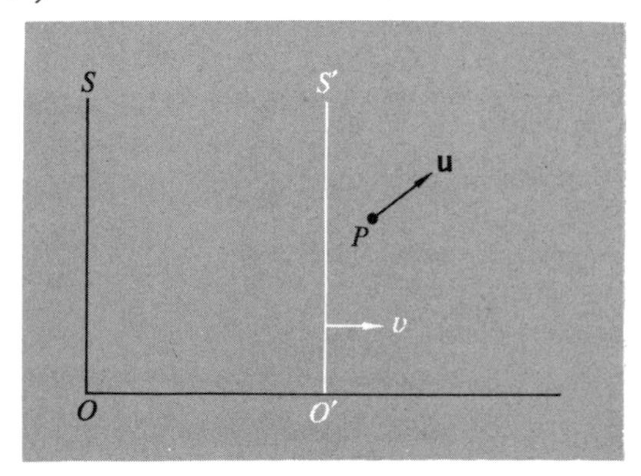

*Fig. 7–2 Particle having an arbitrary velocity **u** with respect to frame S. With the help of the velocity-addition laws, the values of its energy and its momentum components as measured in S and S′ are shown to be linearly related.*

The energy and momentum of the particle in the two frames are as follows:

In S:

$$\left.\begin{aligned} E &= \gamma(u)m_0c^2 \\ p_x &= \gamma(u)m_0u_x \\ p_y &= \gamma(u)m_0u_y \end{aligned}\right\} \quad \text{where } \gamma(u) = (1 - u^2/c^2)^{-1/2} \tag{7–3}$$

In S':

$$\left.\begin{aligned} E' &= \gamma(u')m_0c^2 \\ p_x' &= \gamma(u')m_0u_x' \\ p_y' &= \gamma(u')m_0u_y' \end{aligned}\right\} \quad \text{where } \gamma(u') = (1 - u'^2/c^2)^{-1/2} \tag{7–4}$$

The one big step in relating these two sets of dynamical quantities is to express $\gamma(u)$ in terms of quantities measured in S', or $\gamma(u')$ in terms of quantities measured in S. Let us take the latter. We have

$$\begin{aligned} \gamma(u') &= [1 - (u')^2/c^2]^{-1/2} \\ &= [1 - (u_x')^2/c^2 - (u_y')^2/c^2]^{-1/2} \end{aligned} \tag{7–5}$$

We shall treat this by easy stages. First, consider the following:

$$\begin{aligned} 1 - (u_x')^2/c^2 &= 1 - \frac{(u_x - v)^2}{c^2(1 - vu_x/c^2)^2} \\ &= \frac{(1 - vu_x/c^2)^2 - (u_x - v)^2/c^2}{(1 - vu_x/c^2)^2} \\ &= \frac{1 - u_x^2/c^2 - v^2/c^2 + (vu_x/c^2)^2}{(1 - vu_x/c^2)^2} \end{aligned}$$

Therefore,

$$1 - (u_x')^2/c^2 = \frac{(1 - u_x^2/c^2)(1 - v^2/c^2)}{(1 - vu_x/c^2)^2} \tag{7–6a}$$

Next, note that, from equations (7–2), we have

$$(u_y')^2/c^2 = \frac{(u_y^2/c^2)(1 - v^2/c^2)}{(1 - vu_x/c^2)^2} \tag{7–6b}$$

Subtracting Eq. (7–6b) from (7–6a), we get

$$1 - (u')^2/c^2 = \frac{(1 - u^2/c^2)(1 - v^2/c^2)}{(1 - vu_x/c^2)^2}$$

in which we recognize the squares of the reciprocals of $\gamma(u')$, $\gamma(u)$, and $\gamma(v)$.

We have, in fact,

$$\gamma(u') = \gamma(v)\gamma(u)(1 - vu_x/c^2) \tag{7–7}$$

Now taking this result in conjunction with the first of equations (7–4), we have

$$E' = \gamma(v)[\gamma(u)m_0c^2 - v\gamma(u)m_0u_x]$$

which, by reference to equations (7–3), can be expressed as follows:

$$E' = \gamma(v)(E - vp_x) \tag{7–8}$$

Again, taking the equation for p_x', we have

$$p_x' = \gamma(v)\gamma(u)m_0(u_x - v)$$

i.e.,

$$p_x' = \gamma(v)(p_x - vE/c^2) \tag{7–9}$$

Finally, taking the equation for p_y', we find

$$p_y' = \gamma(u)m_0u_y$$

Therefore,

$$p_y' = p_y \tag{7–10}$$

Let us collect together the transformations from S to S' expressed by Eqs. (7–8), (7–9), and (7–10), plus the corresponding transformations from S' to S:

LORENTZ TRANSFORMATIONS
FOR MOMENTUM AND ENERGY

$$\begin{aligned} p_x' &= \gamma(p_x - vE/c^2) \qquad & p_x &= \gamma(p_x' + vE'/c^2) \\ p_y' &= p_y & p_y &= p_y' \\ p_z' &= p_z & p_z &= p_z' \\ E' &= \gamma(E - vp_x) & E &= \gamma(E' + vp_x') \end{aligned}$$

with $\gamma = (1 - v^2/c^2)^{-1/2}$, where v is the velocity of S' as measured in S

(7–11)

One striking feature of equations (7–11) is that the momentum components and the energy appear in *linear* combinations only; no such simple connection exists in Newtonian mechanics. This linearity is not merely interesting; it has consequences of great value. For, although we obtained the equations in terms of a single particle, there is nothing to stop us from taking E and $\mathbf{p}$ as the total energy and the total momentum of a whole collection of noninteracting particles with arbitrary velocities. It is just a matter of writing down equations like (7–8), (7–9) and (7–10) for each particle separately and adding them all up. Thus equations (7–11), in turn, hold good for any system of particles, and so (as we argued more informally in the last section) does the invariance expressed by Eq. (7–1).

An example. Elastic scattering of identical particles. Let us see how the energy-momentum transformations can be applied to the problem, already discussed in Chapter 6, of the symmetrical elastic collision of two identical particles—e.g., two protons. Figure 7–3 shows such a collision as observed first in the zero-momentum frame S', and then in the laboratory frame S (in which one of the particles is assumed to be initially stationary). As observed in S', each particle has a momentum of magnitude p' both before and after the collision, but the collision turns each momentum vector through 90°. The velocity v relating S' and S is defined by Eq. (6–20):

$$v = \frac{c^2 p'}{E'} \tag{7–12}$$

which expresses the general connection, as we first propounded it in Eq. (1–13), relating velocity, momentum, and energy for any particle. We can use it here because v is not only the relative velocity of S and S', but also the speed, as measured in S', of

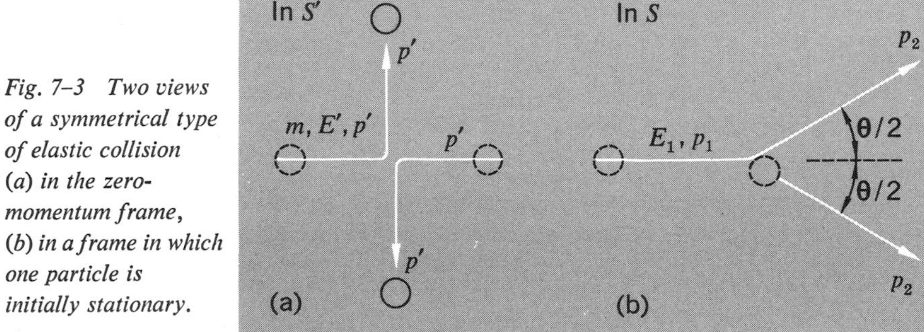

Fig. 7–3 Two views of a symmetrical type of elastic collision (a) in the zero-momentum frame, (b) in a frame in which one particle is initially stationary.

the particle that was initially stationary in S. (It follows, from the symmetry of this particular problem, that it is also the speed in S' of the other particle.)

Let us consider the state of affairs for each particle *after* the collision.

In S':

$$\left.\begin{aligned} p_x' &= 0 \\ p_y' &= p' \\ E' &= mc^2 = \gamma(v)m_0c^2 \end{aligned}\right\} \tag{7–13}$$

Using the equations for transforming momentum components between different frames, we have:

In S:

$$p_x = \gamma(v)(p_x' + vE'/c^2)$$

In this expression we put $p_x' = 0$, and [using Eq. (7–12)] we have $vE'/c^2 = p'$. Thus

$$p_x = \gamma(v)p'$$
$$p_y = p_y' = p'$$

Therefore,

$$\tan\frac{\theta}{2} = \frac{p_y}{p_x} = \frac{1}{\gamma(v)} \tag{7–14}$$

where θ is the angle between the proton directions after scattering, as observed in the laboratory frame [Fig. 7–3(b)].

Clearly $\tan(\theta/2) < 1$, so $\theta/2 < 45°$, $\theta < 90°$. To get $\gamma(v)$ in terms of the initial total energy E_1 of the incident proton (as measured in the laboratory), we can again make use of the energy-momentum invariant of Eq. (7–1):

$$\begin{aligned} (2mc^2)^2 &= (E_1 + m_0c^2)^2 - (cp_1)^2 \\ &= [E_1^2 - (cp_1)^2] + 2E_1m_0c^2 + (m_0c^2)^2 \\ &= (m_0c^2)^2 + 2E_1m_0c^2 + (m_0c^2)^2 \end{aligned}$$

Therefore,

$$\left(\frac{m}{m_0}\right)^2 = \frac{E_1 + m_0c^2}{2m_0c^2}$$

$$\gamma(v) = \left(\frac{E_1 + m_0c^2}{2m_0c^2}\right)^{1/2} \tag{7–15}$$

It is then a straightforward matter to relate Eq. (7–14) to the rather different statement of the result as given in Eq. (6–31).

You will see that the array of equations (7–11) bears a striking resemblance to the set of Lorentz transformations for space and time [cf. equations (3–16)]. Taking the momentum components to fill a role analogous to that of the position coordinates, then the comparison of equations (7–11) with the original Lorentz transformations shows that E/c^2 is the quantity analogous to t. We can say, in fact, that the three components of a linear momentum vector transform like the three components of a position vector, and the total energy (a scalar quantity) transforms like the time. The invariance of the combination $E^2 - (cp)^2$, as expressed in Eq. (7–1), is one immediate consequence of this. The rest energy E_0 is the invariant of the energy-momentum transformations, just as the space-time interval s is the invariant of the Lorentz transformations proper.

In Newtonian mechanics we are accustomed to thinking of measures of space as being definable without reference to time, and vice versa. Likewise, we are accustomed to thinking of momentum and energy as representing essentially different (although to some extent related) properties of a body. We have now seen how these distinctions, both kinematic and dynamic, are blurred in special relativity. The specification of time in one system involves both position and time in another system; the specification of energy involves both energy and momentum in another system. Because of this, and without being in any sense metaphysical, it is entirely appropriate to enlarge the framework of our formal description of things by thinking of a single four-dimensional space-time structure, rather than of a three-dimensional spatial structure that takes on different appearances at different times. In Chapter 3 we alluded briefly to this unification in the description of point events. We shall now express it more formally by saying that the *kinematic* state of a particle is expressible by a single 4-vector whose components are (x, y, z, ict) and whose length is $s\sqrt{-1}$ measured in any frame:

$$-s^2 = x^2 + y^2 + z^2 + (ict)^2$$

Likewise, the *dynamical* state of a particle is expressible by a single 4-vector whose components are $(p_x, p_y, p_z, iE/c)$ and whose "length" (in momentum units) is iE_0/c as measured in any frame:

$$-E_0{}^2/c^2 = p_x{}^2 + p_y{}^2 + p_z{}^2 + (iE/c)^2$$

The Lorentz transformations can then be regarded as a prescription for transforming the various components of a 4-vector from one set of axes to another—often described as the "mapping" of the vector onto various coordinate systems in the four-dimensional world. This way of representing the relativistic scheme of things is attractive in a formal sense, and can also be very useful if one has once learned to exploit it—which primarily means developing the appropriate fluency in matrix algebra. We shall not go any further with it here, however, since it is not essential and really adds nothing to the basic physics of relativity.

FORCE IN RELATIVISTIC MECHANICS

In contrast to the Newtonian conception, it is easy to show that in relativity the quantity force, in general, is not codirectional with the acceleration it produces . . . It is also easy to show that these force components have no simple transformation properties . . . All these modifications, important as they are from the mathematical point of view, do not radically affect the conception of force. Yet an important point should be noted: on grounds of the rejection of an absolute simultaneity of two distant events, special relativity comes to the conclusion that action at a distance has to be excluded as a legitimate physical notion.

Max Jammer, *Concepts of Force* (1957)

In all of our discussion of relativistic dynamics so far, we have placed an almost exclusive emphasis on the use of the energy and momentum conservation laws for an isolated system of particles. We have tried to give some feeling for the variety of problems that can be discussed in these terms. But when all is said and done, this approach is not always the most convenient or useful. Many, many problems in dynamics can best be treated (and perhaps can only be efficiently treated) in terms of the motion of particles under the action of a given set of forces. Take the Rutherford scattering of α particles, for example. Granted that the force becomes vanishingly small at large separations between two particles (and it is this assumption that underlies all the calculations we have done in Chapter 6 and in this chapter), one can make perfectly correct statements about the relation between final directions and velocities of the colliding nuclei. But this does not tell us the probability that an α particle will in fact be deflected through a certain angle. Only when we put in the explicit law of force do we find answers to such questions as these. And the discovery and specification of laws of force is a central concern of physics. It is certainly important, therefore, to know how to transform forces and equations of motion so as to give a description of them from the point of view of different inertial frames. Since in special relativity the acceleration is not an invariant, we know that we cannot enjoy the simplicity of Newtonian mechanics, but we can certainly arrive at some useful and meaningful statements.

The starting point, which we indeed made use of in the initial stages of our approach to relativity (see Chapter 1) is Newton's law in the form

$$\mathbf{F} = \frac{d\mathbf{p}}{dt} = \frac{d}{dt}(m\mathbf{v}) \qquad \text{where } m = m_0(1 - v^2/c^2)^{-1/2} \tag{7–16}$$

We take this as a definition of **F**. It is a natural extension (and the simplest extension) of the nonrelativistic result. It is not a statement that can be independently proved. On the other hand, if the analytical form of **F** is given, in terms of coordinates, velocities, etc., we must certainly demand that the left and right sides of Eq. (7–16) transform in the same way under Lorentz transformations. Assuming that this necessary condition has been met, the transformations of the components of $d\mathbf{p}/dt$ tell us how force components transform in special relativity.

Our approach to the problem will be as follows: At any instant a particle has a well-defined velocity **v** as measured in a laboratory frame of reference. We can picture the particle as being instantaneously in a rest frame that has this velocity **v** with respect to the laboratory. We shall imagine that, as measured in the rest frame, a force F_{0x} is applied parallel to **v**, causing an acceleration a_{0x}. The mass as measured in this frame is just the rest mass m_0. Hence we have

$$F_{0x} = m_0 a_{0x} \tag{7–17}$$

Now in the laboratory frame we have a momentum given by

$$p_x = \gamma m_0 v = \frac{m_0 v}{(1 - v^2/c^2)^{1/2}}$$

and hence we judge the force to be F_x, where

$$F_x = \frac{dp_x}{dt} = \frac{m_0}{(1 - v^2/c^2)^{1/2}}\frac{dv}{dt} + m_0 v \frac{d}{dt}[(1 - v^2/c^2)^{-1/2}]$$

If we put $dv/dt = a_x$ (the observed acceleration in the laboratory) we have

$$F_x = \frac{m_0 a_x}{(1 - v^2/c^2)^{1/2}} + \frac{m_0(v^2/c^2)a_x}{(1 - v^2/c^2)^{3/2}}$$

which, when we collect terms, simplifies to

$$F_x = \gamma^3 m_0 a_x \tag{7–18}$$

There is, however, a very simple connection between a_x and a_{0x}:

$$a_x = \frac{1}{\gamma^3} a_{0x} \tag{7–19}$$

This is the particular case of the transformation of accelerations along x for $u_{0x} = 0$ [cf. Eq. (5–24)]. Thus Eq. (7–18) can be written

$$F_x = \gamma^3 m_0 \frac{a_{0x}}{\gamma^3} = m_0 a_{0x}$$

i.e.,

$$F_x = F_{0x} \tag{7–20}$$

This is a striking result. Despite the change of the measures of mass and acceleration in the two frames, the measure of the x component of force remains the same.

When we make a similar calculation for the transverse force, we find that this invariance does not hold. In the instantaneous rest frame we have

$$F_{0y} = m_0 a_{0y} \tag{7–21}$$

In the laboratory frame, the force F_y applied perpendicular to the momentum vector $m\mathbf{v}$ will, during some very brief interval of time, leave the magnitude of the velocity unaltered; it merely changes the direction of $\mathbf{v}$ slightly by introducing a small transverse component. Thus to a good approximation (which becomes perfect in the limit $\Delta t \to 0$) the mass remains unchanged at γm_0, and the transverse impulse can be written

$$F_y \, \Delta t = \gamma m_0 \, \Delta v_y$$

Hence we have, in this case,

$$F_y = \gamma m_0 a_y \tag{7–22}$$

Again there is a simple relation between the accelerations in the two frames, given that one of them is measured in the rest frame:

$$a_y = \frac{1}{\gamma^2} a_{0y} \tag{7–23}$$

this being a special case of Eq. (5–26). We therefore have

$$F_y = \gamma m_0 \frac{a_{0y}}{\gamma^2} = \frac{1}{\gamma} m_0 a_{0y}$$

i.e.,

$$F_y = \frac{1}{\gamma} F_{0y} \tag{7–24}$$

In the above results one can discern the feature (mentioned in the quotation at the head of this section) that in general force and acceleration are not parallel vectors. Combining Eqs. (7–18) and (7–22) we have

$$\frac{F_y}{F_x} = \frac{1}{\gamma^2} \frac{a_y}{a_x}$$

Only in the instantaneous rest frame of a body ($\gamma = 1$) can one guarantee that **F**, as defined by the time derivative of momentum, is in the same direction as the acceleration.

It is perhaps worth pointing out that the special cases of force transformation represented by Eqs. (7–20) and (7–24) can be derived in simple physical terms.[1] For the x transformation, we can consider the work done by the force, and the resulting increase of energy as manifested in a mass increase:

$$\Delta E = F_x \Delta x = c^2 \Delta m$$

where

$$\Delta x = v \Delta t$$

$$\Delta m = \Delta\left[\frac{m_0}{(1 - v^2/c^2)^{1/2}}\right] = \frac{m_0 v \,\Delta v}{(1 - v^2/c^2)^{3/2} c^2}$$

These at once give us

$$F_x = \gamma^3 m_0 a_x$$

which reproduces Eq. (7–18).

To relate the acceleration a_x to the acceleration a_{0x} as measured in the instantaneous rest frame of the particle, we take the equations for uniformly accelerated motion:

[1]See, for example, W. P. Ganley, *Am. J. Phys.*, **31**, 510–516 (1963) for a nice discussion.

At time t:

$$x = x_1, \qquad \text{(say)}$$

At time $t + \Delta t$:

$$x + \Delta x = x_1 + v\,\Delta t + \tfrac{1}{2}a_x(\Delta t)^2$$

The space and time coordinates (x, t) and $(x + \Delta x, t + \Delta t)$ define two events as observed in S. Let us obtain the coordinates of these same events as observed in the rest frame S' of the particle. To do this we use the Lorentz transformations:

$$x_0 = \gamma(x - vt)$$
$$t_0 = \gamma(t - vx/c^2)$$

Applying these to the two events in turn, we have the following:

First event:

$$x_0 = \gamma(x_1 - vt)$$
$$t_0 = \gamma(t - vx_1/c^2)$$

Second event:

$$\begin{aligned} x_0 + \Delta x_0 &= \gamma[x_1 + v\,\Delta t + \tfrac{1}{2}a_x(\Delta t)^2 - v(t + \Delta t)] \\ &= \gamma[x_1 + \tfrac{1}{2}a_x(\Delta t)^2 - vt] \\ t_0 + \Delta t_0 &= \gamma[t + \Delta t - (v/c^2)\{x_1 + v\,\Delta t + \tfrac{1}{2}a_x(\Delta t)^2\}] \end{aligned}$$

Subtracting,

$$\Delta x_0 = \gamma[\tfrac{1}{2}a_x(\Delta t)^2] \tag{7–25}$$
$$\Delta t_0 = \gamma[(1 - v^2/c^2)\,\Delta t - \tfrac{1}{2}(va_x/c^2)(\Delta t)^2]$$

If Δt is sufficiently short, the second term in the equation for Δt_0 becomes negligible compared to the first, and we have

$$\Delta t_0 \approx \gamma(1 - v^2/c^2)\,\Delta t = \Delta t/\gamma$$

Substituting $\Delta t = \gamma\,\Delta t_0$ in Eq. (7–25) gives us

$$\Delta x_0 = \gamma[\tfrac{1}{2}a_x(\gamma\,\Delta t_0)^2]$$

or

$$\Delta x_0 = \tfrac{1}{2}(\gamma^3 a_x)(\Delta t_0)^2$$

But this is the equation of uniformly accelerated motion for a particle initially at rest. Thus the acceleration a_{0x} as measured in the rest frame is given by $a_{0x} = \gamma^3 a_x$, which reproduces Eq. (7–19). Thus we have reproduced both Eqs. (7–18) and (7–19) and can combine them to demonstrate the invariance of F_x. This development may seem unduly long-winded, because we have carefully spelled out the transformations. It can be briefly (though somewhat glibly) summarized by saying that time dilation causes the time Δt_0 in the rest frame to correspond to $\gamma \, \Delta t_0$ in the frame S, and that Lorentz contraction causes Δx_0, the distance traveled in the rest frame in consequence of the acceleration, to correspond to $\Delta x_0/\gamma$ in S. On this basis we would arrive at once at the relation

$$\Delta x_0/\gamma = \tfrac{1}{2} a_x (\gamma \, \Delta t_0)^2$$

However, where nonproper measurements are involved, as they are here, it pays to be methodical and explicit.

To obtain the transformation of F_y, we set up Eq. (7–22) and then argue the transformation of transverse acceleration from the two statements

$$\Delta y = \tfrac{1}{2} a_y (\Delta t)^2$$
$$\Delta y' = \tfrac{1}{2} a_{0y} (\Delta t')^2$$

with $\Delta y = \Delta y'$ and $\Delta t = \gamma \, \Delta t'$. In this case, since the motion takes place at a constant value of x' in the rest frame, we can apply time dilation directly without any qualms.

MAGNETIC ANALYSIS OF RELATIVISTIC PARTICLES

There is one direct application of force laws in relativistic motions that we could not possibly omit from a chapter having so much to do with atomic particles. This is the deviation of a charged particle in a magnetic field. It provides one of the chief diagnostic tools in particle physics, because it reveals both the sign of the charge and the magnitude of the momentum of a particle. The bubble-chamber photograph in Fig. 7–4 is a beautiful example of the use of the technique. The basis of it is the fact that a moving charge q in a magnetic field $\mathbf{B}$ experiences a transverse force proportional to its velocity, according to the vector force law

$$\mathbf{F} = \text{const.}\ (q\mathbf{v} \times \mathbf{B})$$

In the MKS system of measurement, the value of the constant is

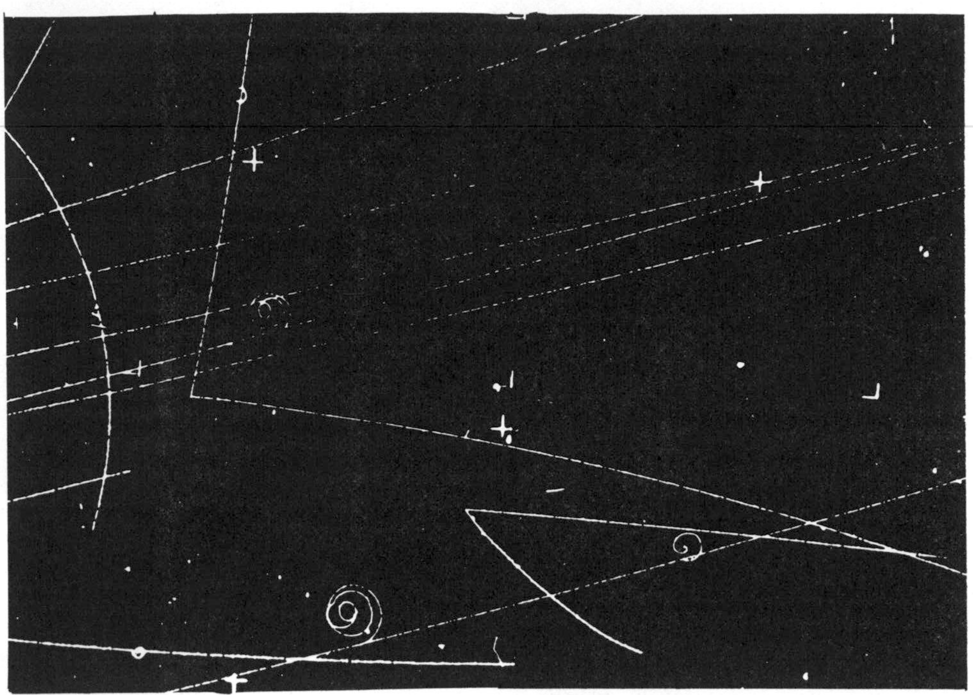

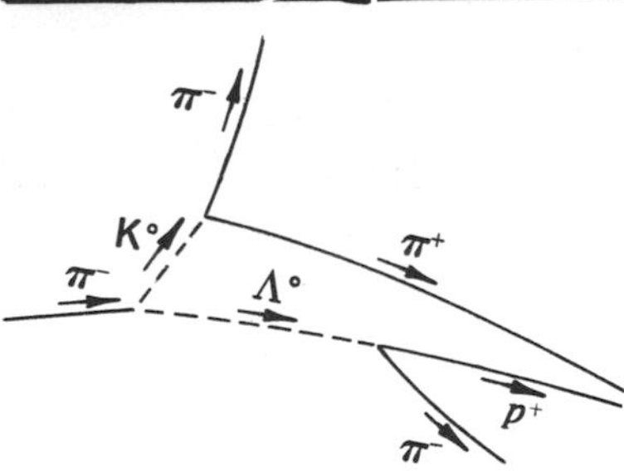

Fig. 7–4 Hydrogen bubble-chamber photograph of the production of two unstable neutral particles (K^0 and Λ^0) by the collision of a π^- meson with a proton in a 500-liter liquid hydrogen bubble chamber developed by Prof. L. W. Alvarez and his group at Berkeley. (Photo courtesy of Prof. Alvarez and Lawrence Radiation Laboratory, Berkeley, Calif.) The π^- enters from the left and its track ends abruptly, marking the point of origin of the particles K^0 and Λ^0. These subsequently decay into pairs of charged particles ($K^0 \to \pi^+ + \pi^-$, $\Lambda^0 \to \pi^- + p$), giving two pairs of forked tracks. Analysis of the tracks of the charged particles in each fork shows that the total linear momentum vector in each fork is directed outward from the point where the initial interaction took place. This picture was used as an example by Prof. D. A. Glaser, inventor of the bubble chamber, in his Nobel lecture in 1960. (See Nobel Lectures, Physics, 1942–1962, Elsevier, 1964.)

unity, by definition. If the field direction and the particle velocity are perpendicular, the motion of the particle remains in a plane perpendicular to the magnetic field. At every instant, the force exerted on the particle is at right angles to the momentum vector **p**, and the magnitude of the force is given by

$$F = qvB \qquad \text{(MKS system)}$$

Thus in a short time Δt the particle acquires a transverse momentum given by

$$\Delta p = F\,\Delta t = qvB\,\Delta t$$

This means that the momentum vector is turned through a small angle $\Delta\theta$ such that

$$\Delta\theta = \frac{\Delta p}{p} = \frac{qvB}{p}\Delta t \qquad (7\text{–}26)$$

Hence the velocity vector has turned through $\Delta\theta$ also, in a time during which the particle has traveled a distance Δs such that

$$\Delta s = v\,\Delta t$$

But Δs and $\Delta\theta$ define a radius of curvature R, and in fact we have

$$\Delta\theta = \frac{\Delta s}{R} = \frac{v\,\Delta t}{R} \qquad (7\text{–}27)$$

Combining Eqs. (7–26) and (7–27) we have

$$p = qBR \qquad \text{(MKS units)} \qquad (7\text{–}28)$$

It is interesting to note that this is precisely the same result as one gets from Newtonian mechanics, except that in Eq. (7–28) one must remember that $p = \gamma m_0 v$, not $m_0 v$ simply. In very many cases one can be sure that $q = \pm e$, so that a knowledge of B and a measurement of the radius of curvature of the path will determine the momentum of the particle. Figure 7–4 shows a whole collection of such motions, and exemplifies the wealth of information that may be revealed by a single picture with the magnificent techniques that have been developed in particle physics.

GENERAL FORCE TRANSFORMATIONS; ACTION AND REACTION

In the section before last we considered some cases of force transformation that were quite special, because one of the two frames chosen was the instantaneous rest frame of the particle. But we can extend the analysis so as to yield transformations of force between any two frames, through the definitions

$$\mathbf{F} = \frac{d\mathbf{p}}{dt} \qquad \mathbf{F}' = \frac{d\mathbf{p}'}{dt'}$$

The calculation involves the use of the relativistic transformations for velocity, momentum, and energy (as well as for x and t), so for convenience we shall restate here a minimal number of the formulas needed to obtain the two basic forms of the transformation—for force components either parallel or perpendicular to the direction of relative motion of two reference frames.

Let the momentum and the energy of a particle, as measured in S, be $\mathbf{p}$ and E at the space-time point (x, y, z, t), and let $\mathbf{p}'$ and E' represent the momentum and the energy of the particle as measured in S' at the *same* space-time point (x', y', z', t'). Then we have

$$\left.\begin{aligned} x' &= \gamma(x - vt) \\ y' &= y \\ t' &= \gamma(t - vx/c^2) \end{aligned}\right\} \tag{7–29}$$

$$\left.\begin{aligned} u_x' &= \frac{u_x - v}{1 - vu_x/c^2} \\ u_y' &= \frac{u_y/\gamma}{1 - vu_x/c^2} \end{aligned}\right\} \tag{7–30}$$

$$\left.\begin{aligned} p_x' &= \gamma(p_x - vE/c^2) \\ p_y' &= p_y \\ E' &= \gamma(E - vp_x) \end{aligned}\right\} \tag{7–31}$$

Since the force $\mathbf{F}'$ on a particle, as measured in S', is defined by

$$\mathbf{F}' = d\mathbf{p}'/dt'$$

it follows from equations (7–31) and the Lorentz transformation for time that

$$F_x' = \frac{dp_x'}{dt'} = \frac{\dfrac{dp_x'}{dt}}{\dfrac{dt'}{dt}} = \frac{\gamma\left(\dfrac{dp_x}{dt} - \dfrac{v}{c^2}\dfrac{dE}{dt}\right)}{\gamma\left(1 - \dfrac{v}{c^2}\dfrac{dx}{dt}\right)}$$

i.e.,

$$F_x' = \frac{F_x - (v/c^2)\, dE/dt}{1 - vu_x/c^2} \tag{7–32}$$

Now dE/dt is the rate of change of the particle's energy as measured in S. In Newtonian mechanics we could immediately identify this with the quantity $\mathbf{F} \cdot \mathbf{u}$, the rate at which the force $\mathbf{F}$ does work. This also holds good in relativistic dynamics, as we can see by the following argument:

We have

$$E^2 = c^2p^2 + E_0{}^2 = c^2(\mathbf{p}\cdot\mathbf{p}) + E_0{}^2$$

Therefore,

$$\begin{aligned} E\,dE/dt &= c^2\mathbf{p}\cdot(d\mathbf{p}/dt) \\ &= c^2\mathbf{p}\cdot\mathbf{F} \end{aligned}$$

But $E = mc^2$. Therefore,

$$dE/dt = \mathbf{F}\cdot(\mathbf{p}/m) = \mathbf{F}\cdot\mathbf{u}$$

Hence Eq. (7–32) becomes

$$F_x' = \frac{F_x - (v/c^2)(\mathbf{F}\cdot\mathbf{u})}{1 - vu_x/c^2} \tag{7–33}$$

Similarly (but much more simply), we find

$$F_y' = \frac{F_y}{\gamma(1 - vu_x/c^2)} \tag{7–34}$$

(Note that, if $\mathbf{u} = 0$, we have once again the simple results $F_x' = F_x$, $F_y' = F_y/\gamma$.) An equation exactly like Eq. (7–34) holds for the relation between F_z' and F_z, and the expressions for F_x, F_y, F_z in terms of S' measurements involve writing equations equivalent to Eqs. (7–33) and (7–34) with the sign of v reversed.

Equation (7–33) is a very interesting one, in that it tells us that the measure of a force in one frame involves the measure of the *power* developed by the force in another frame. It is one more manifestation of the intermingling of space and time measurements inherent in the relativistic description of things, and has no counterpart in classical mechanics. It has been commented on as follows[1]:

> In the classical mechanics there have always been two strains of thought. The two aspects of "force" as "the time rate of change of momentum," and as "the space rate of change of energy," have with different writers been given different degrees of prominence. Galileo developed the former, Huyghens the latter. In the light of four-dimensional vectors the two ideas become unified, and differ only as partial aspects of a greater concept . . . .

Pursuing this last comment, we may note that in a four-dimen-

[1]By E. Cunningham, in an old but excellent book, *Relativity, The Electron Theory and Gravitation*, Longmans, Green, London, 1921.

sional space-time world, the force components F_x, F_y, and F_z represent only three of the components of some 4-vector. What is the fourth component, and what is the vector? The clue is already provided by Eq. (7–33), in which the quantity $\mathbf{F} \cdot \mathbf{u}$ appears. A simple calculation, much like that for the transformation of F_x', leads to the following result:

$$\mathbf{F}' \cdot \mathbf{u}' = \frac{(\mathbf{F} \cdot \mathbf{u}) - vF_x}{1 - vu_x/c^2} \tag{7–35}$$

The quantity $(\mathbf{F} \cdot \mathbf{u})/c$ has the same dimensions as the space components of $\mathbf{F}$ and one can, with a little extra juggling, construct an invariant from these four quantities. The exercise is a rather artificial one, however, and only serves to emphasize how momentum and energy, rather than force, provide the foundation of relativistic dynamics.

It is worth pointing out that one of Newton's basic assertions about forces between bodies—the equality of action and reaction—has almost no place in relativistic mechanics. It must essentially be a statement about the forces acting on two bodies, as a result of their mutual interaction, *at a given instant*. And, because of the relativity of simultaneity, this phrase has no unique meaning unless the points at which the forces are applied are separated by a negligible distance. It was in this sense that Max Jammer, in the remarks quoted earlier in this chapter, asserted that the concept of action at a distance has no place in relativistic dynamics. Even if the force on one object is known to be solely due to the presence of some other object, we have no unique way of describing their *mutual* interaction; we can only describe the force exerted on either body, separately, at some given point in space-time. This does *not* imply that we can no longer write down a quantitative statement of the force exerted on one body by another, as described in a given reference frame.[1] What the relativistic analysis does do, however, is to compel us to conclude that, according to measurements in a given inertial frame, the forces of action and reaction are in general *not* equal and opposite, and so the total momentum of the interacting particles is not conserved, instant by instant. This fact leads, if one wishes to hold to conservation of momentum, to the idea that momentum (as well as energy) may reside in the field that describes the interaction of separated particles. As far as the particles alone are concerned, conservation of momentum applies only when

[1]We shall, in fact, be doing this repeatedly in Chapter 8.

one compares the initial and final situations (before the interaction begins and after it has ceased). During the interaction itself, the momentum of the interaction field must be brought in if one is to have total momentum conservation at all times in all frames of reference.

It is another consequence of the force transformations that if, as measured in one frame, the force on a body depends on its position but not on its velocity, then as measured in other frames the force depends on the body's velocity as well as on its position. Probably the most important example of this is in electromagnetism. The force on a moving electron due to stationary charges is given simply by Coulomb's law. But if we imagine ourselves in a frame in which the charges that caused the force are moving, then the force on the electron depends on its velocity as well as on the motion of the other charges. It involves, in fact, the magnetic force between what are effectively two currents. In this result we see the germ of the development by which electric and magnetic fields can be shown to be intimately related. To do it justice, however, this subject needs a full discussion on its own account; the hint that we have just given is clearly quite inadequate. The final chapter of this book is therefore devoted entirely to a discussion of the elements of this fascinating development.

PROBLEMS

7–1 A K meson traveling through the laboratory breaks up into two π mesons. One of the π mesons is left at rest. What was the energy of the K? What is the energy of the remaining π meson? (Rest mass of K meson = 494 MeV; rest mass of π meson $\simeq$ 137 MeV.)

7–2 An electron-positron pair can be produced by a γ ray striking a stationary electron:

$$\gamma + e^- \rightarrow e^- + e^+ + e^-$$

What is the minimum γ-ray energy that will make this process go?

7–3 Suppose that a certain accelerator can give protons a kinetic energy of 200 GeV. The rest mass m_0 of a proton is 0.938 GeV. Calculate the largest possible rest mass M_0 of a particle X that could be produced by the impact of one of these high-energy protons on a stationary proton in the following process:

$$p + p \rightarrow p + p + X$$

7–4 A particle of rest mass M collides elastically with a stationary particle of rest mass $m \ll M$. If the initial velocity of M is such that $\gamma = M/m$, show that the maximum angle through which M can be scattered is approximately $m/\sqrt{3}M$. Show also that the maximum recoil angle of m is 90°.

7–5 A particle of rest mass m and velocity v collides elastically with a stationary particle of rest mass M. Express the recoil and scattering angles in terms of the corresponding angles in the zero-momentum system. Show that your answers reduce to the nonrelativistic ones if $v \ll c$.

7–6 The kinetic energy K of a system in the lab frame is related to the kinetic energy K^* in the center-of-mass frame in the nonrelativistic case by the expression $K = K^* + MV^2/2$, where M is the total mass of the system and V is the velocity of the center of mass. What is the analogous expression for the relativistic case? Show that it reduces to the above result if all speeds are much less than c.

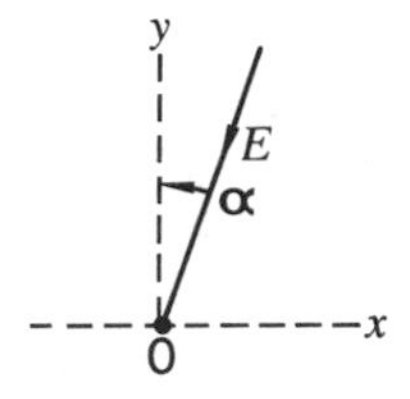

7–7 (a) A photon of energy E travels toward the origin O of a coordinate system S, making an angle α with the y axis (see the figure). Using the Lorentz transformations for momentum and energy [equations (7–11)] calculate the energy of the photon in a frame S' in which it is observed to travel straight down the y' axis.

(b) Apply the same kind of analysis to the Mössbauer problem (Problem 6–6) by calculating the energy of the received photon in the frame in which the absorber on the rim of the wheel is instantaneously at rest. Note that this gives us an alternative method for deriving the Doppler-effect formula.

7–8 Problem 5–18 asked you to analyze the problem of an object that travels for a time t (earth time) at a constant acceleration of 9.8 m/sec^2 (g) in its own rest frame. Solve this problem again, using the result $F_x = F_x'$ together with the variable time-dilation factor $\gamma(v)$ that relates elements of time dt and dt' at any given stage.

7–9 The cyclotron is a machine used to accelerate positive ions to energies of a few MeV (see the figure). The dees are metal shells which shield out electric fields but not magnetic ones. There is thus effectively

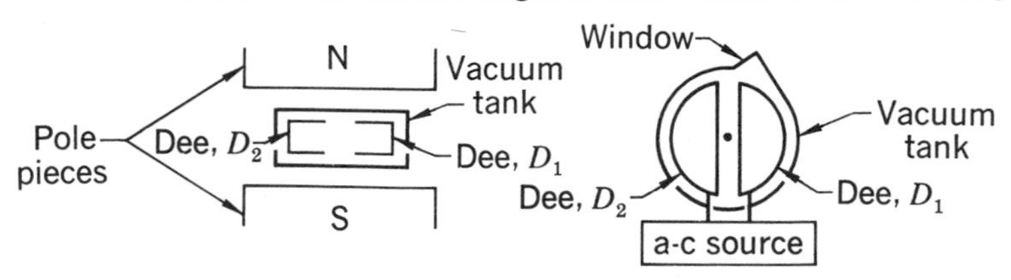

Vertical cross section of cyclotron   Horizontal cross section

a constant magnetic field B throughout the vacuum tank and an alternating (sinusoidal) electric field in the small gap between the dees. If this field is in phase with the ions, they receive a "kick" from it every half-cycle. These "kicks" increase the energy, and hence the radius of the orbit of the ions until they eventually exit through the window.

(a) What is the frequency at which the ions circle the cyclotron? Show that your answer is independent of the kinetic energy of the ions if $v \ll c$. Evaluate your result numerically for protons if $B =$ 1.5 MKS units (webers/m^2).

(b) If the maximum usable diameter inside the dees is 60 cm, what are the maximum kinetic energy and velocity such a proton can attain?

(c) If the protons receive a "kick" of 10^5 volts every time they cross the gap, how many revolutions do they make before leaving through the window? How long does the whole process take?

(d) What technical difficulties arise if the maximum usable diameter is greatly increased (to 6 m, for example)?

7–10 In a laboratory bubble-chamber experiment [H. Whiteside, J. N. Palmieri, and R. A. Burnstein, *Am. J. Phys.*, **34,** 1005 (1966)], a K^- meson was observed to interact with a stationary proton, yielding a π^+ meson and an unknown (X) particle, giving a set of three tracks as shown in the figure. The magnetic field inside the bubble chamber was 1.70 ± 0.07 webers/m^2 (MKS units).

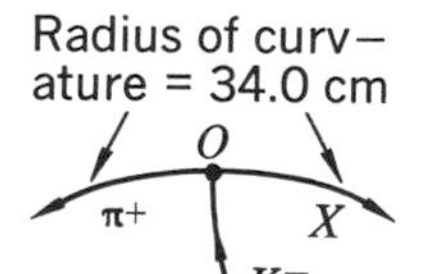

(a) Why do the tracks that originate at the vertex O (the interaction point) indicate that the K^- was at rest at the time of the interaction?

(b) Using the table below, identify the unknown particle.

Name and symbol		*Rest mass, MeV*	*Charge, units of e*
Positron, electron	e^+, e^-	0.511	±1
Muon	μ^+, μ^-	105.7	±1
Pi meson	π^+, π^-	139.6	±1
K meson	K^+, K^-	493.8	±1
Proton	p^+	938.3	1
Neutron	n	939.6	0
Lambda	Λ^0	1115.4	0
Sigma plus	Σ^+	1189.4	1
Sigma zero	Σ^0	1192.3	0
Sigma minus	Σ^-	1197.2	-1
Xi zero	Ξ^0	1314.3	0
Xi minus	Ξ^-	1320.8	-1
Omega minus	Ω^-	1675	-1

What led me more or less directly to the special theory of relativity was the conviction that the electromotive force acting on a body in motion in a magnetic field was nothing else but an electric field.

ALBERT EINSTEIN (1952)

From a letter to the Michelson Commemorative Meeting of the Cleveland Physics Society as quoted by R. S. Shankland, *Am. J. Phys.*, **32,** 16 (1964), p. 35.

8

Relativity and electricity

BY THE END of the 19th century, the basic phenomena of classical electricity and magnetism were well known. Electrically charged bodies exerted forces upon each other, and the effects could be described in terms of electric fields. Electrically *neutral* wires, in which currents flowed, exerted forces upon each other and upon magnets, and the effects could be described in terms of magnetic fields. A time-varying magnetic field could cause a current to flow in a conducting circuit—the phenomenon of *electromagnetic induction.* And electrical disturbances (e.g., radio waves) could radiate out into space. James Clerk Maxwell had, by 1864, developed a set of equations which correctly described the relationships among electric fields, magnetic fields, and stationary and moving electric charges. The relativity principle then demanded that these relationships remain exactly the same no matter what inertial coordinate system the observations were referred to.

The relativity principle imposes severe conditions upon any theory. Suppose, working in an inertial coordinate system S, we discover the law which describes the electric field produced by a moving charge. In a different inertial coordinate system S', the same charge will be observed to be moving with a different velocity, and the electric field may have different values. If we know how to carry out the transformations that relate measurements in S to measurements in S', then we can find the form of the law that describes, in S', what connection exists there between a moving charge and its electric field. The relativity principle requires that the law be exactly the same as that found in S.

The relativity principle may be used in different ways. If we think we know a law of nature which connects various observed quantities, we can use the relativity principle to find out how the observed quantities must transform. If we already know (i.e., think we know) how observed quantities transform, we can test whether or not a theory does satisfy the relativity principle. If not, then something is wrong. Proceeding in this way, Einstein found that Newtonian mechanics did not pass the test, and he was able to propose appropriate modifications.

As we have repeatedly seen in this book, the departures from the predictions of Newtonian mechanics become increasingly greater as the velocities involved approach the velocity of light. This need occasion no surprise. After all, the validity of Newton's mechanics had originally been tested only at low velocities, and it is always risky to apply a law outside its validated range.

What *is* surprising—indeed astonishing—is that Maxwell's laws of electricity and magnetism have required *no* changes at high velocities. The experimental observations available to Maxwell of the effects produced by moving charges were all made on slowly moving charges. Nevertheless, we find today that Maxwell's laws remain accurately correct, even when the velocities of the charges approach very closely the speed of light! In fact, it was Einstein's demand that the Maxwell equations correctly represent electromagnetic phenomena in any inertial coordinate system that enabled him to obtain the transformation expressions for electric and magnetic fields.

Historically, electromagnetic theory preceded special relativity theory—as is attested by the fact that Einstein's great 1905 paper bore the title "On the Electrodynamics of Moving Bodies." But now we have relativity theory, and it stands on its own foundations. It is a powerful tool, and it can be used to develop electromagnetic theory. We shall not undertake anything as ambitious as that in this chapter, but will do enough to illustrate how, by the application of relativistic ideas and techniques, the intimate connection between electric and magnetic phenomena can be demonstrated. And, by way of introduction, we shall first review those results in electromagnetism that will be relevant to our discussion.

COULOMB'S LAW

The experiments of Coulomb and others established that the force which a stationary charge q_1 exerts upon a stationary charge q_2 is directly proportional to the magnitudes of the charges, inversely proportional to the square of the distance between them, and in the direction of the line connecting the two charges. We can write this as follows:

$$\mathbf{F} = k\frac{q_1 q_2}{r^2}\mathbf{e}_r \tag{8–1}$$

where $\mathbf{F}$ is the force exerted by q_1 upon q_2, k is a proportionality factor, r is the magnitude of the distance from q_1 to q_2, and $\mathbf{e}_r$ is a unit vector in the direction from q_1 to q_2.[1] In the CGS system of measurement, the value of the constant k is unity, by definition. In the MKS system it has the value 9×10^9 newton·m^2/coulomb^2 very nearly. The magnitude of the force is inferred from the behavior of q_2—or, to be more precise, from the behavior of the object on which the charge q_2 resides. We shall refer to q_1 as the *source charge* and q_2 as the *test charge*, although, of course, their roles could equally well be reversed.

Since $\mathbf{e}_r$ is equal to $\mathbf{r}/r$, where $\mathbf{r}$ is the vector distance from q_1 to q_2, Coulomb's law can also be written

$$\mathbf{F} = kq_1q_2\frac{\mathbf{r}}{r^3} \tag{8–1a}$$

but it must be appreciated that this is still an inverse-square dependence even though r^3 appears in the denominator.

If a test charge is held stationary between charged plates (for example, in a Millikan apparatus) it experiences a force that can be calculated from Coulomb's law. If the test particle is in motion when it is between the charged plates, as in a cathode-ray tube, experimental measurements of the particle's deflection show that the force on the moving charge is *still* correctly given by Eq. (8–1). This remains true even as the speed of the test charge approaches that of light. The investigation shown in the film

[1]The use of the symbol $\mathbf{e}$ for a unit vector comes from the German word "einheit."

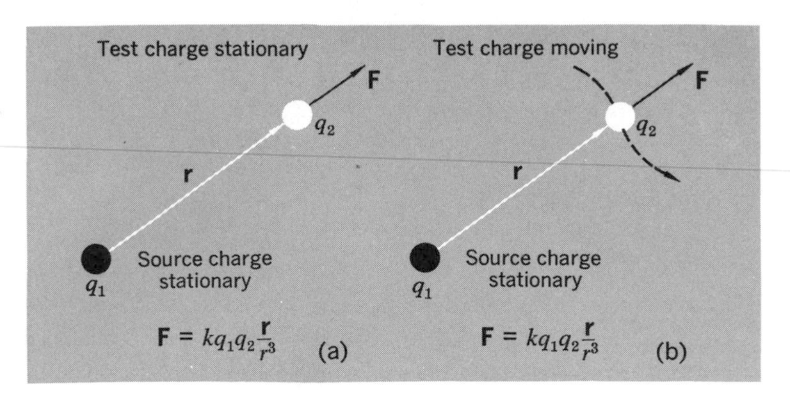

Fig. 8–1 Coulomb's law. (a) As first established, with both charges stationary. (b) It is still true if the test charge is moving, as long as the source charge is at rest.

"The Ultimate Speed" reveals that the energy acquired by a very fast-moving electron when it is acted upon by an electrical force can be calculated by taking the force as correctly given by Eq. (8–1). Generalizing these observations, we can assert that *Coulomb's law correctly gives the force on the test charge, for any velocity of the test charge* (*however high*), *provided the source charge is at rest* (see Fig. 8–1). Since **F** is proportional to the magnitude of the test charge, and independent of its velocity, we can use our knowledge of the force $\mathbf{F}_{elec}$ for a given value of q_2 to define what we call the *electric field* $\boldsymbol{\mathcal{E}}$ at the position of the test charge

$$\mathbf{F}_{elec} = q_2\boldsymbol{\mathcal{E}} \tag{8–2}$$

and so, for the case of a stationary source charge, q_1, we have

$$\boldsymbol{\mathcal{E}} = \frac{kq_1}{r^2}\,\mathbf{e}_r \tag{8–2a}$$

Given the value of $\boldsymbol{\mathcal{E}}$, we can then calculate the Coulomb force on any charged particle at a given position.

THE MAGNETIC FORCE ON A MOVING CHARGE

Despite what we have just said about the Coulomb force on a charged particle being independent of the particle's motion, the fact remains that a charged particle at a given point *may* experience a force that depends on its velocity. Such a force arises if there is at the position of the charge what we call a magnetic field. The field has a well-defined direction, as indicated for example by a compass needle, and the magnetic force on the moving charge is found to be given by the following equation (already cited in Chapter 7):

$$\mathbf{F}_{\text{mag}} = \text{const.}\ (q_2\mathbf{u} \times \mathbf{B})$$

where $\mathbf{u}$ is the velocity of the charged particle (of charge q_2) and $\mathbf{B}$ characterizes the strength and direction of the magnetic field. In the MKS system of measurement, the value of the constant is unity, because of the way in which the units of charge and magnetic field are defined. In the CGS system, the constant turns out to be equal to the numerical magnitude of $1/c$. Why is this? A heavy emphasis on units and systems of measurement would be inappropriate in this chapter. The main interest lies elsewhere. On the other hand, the insight that relativity theory brings to electromagnetism removes much of the seeming arbitrariness of the subject. It exposes the essential relation between the Coulomb law and the magnetic force law, and between the constants that appear in them. And even at the outset it may be helpful to make some further remarks about these two basic force laws.

When electrostatics was first developed, the Coulomb law, Eq. (8–1), was used as a basis for defining the unit of electric charge. Since the CGS system for mechanical measurements was in general use at the time, the unit charge was defined as being of such a magnitude that, if placed 1 cm from a similar charge in vacuum, each charge would exert on the other a force of 1 dyne (= 10^{-5} newton). The force between two arbitrary charges in vacuum could then be simply set equal to q_1q_2/r^2. Thus, using the CGS system, the unit of charge (1 esu) was defined in terms of the force between *stationary* charges.

Much later it was recognized that, since currents are made up of moving charges, an alternative scheme for defining a unit of charge could be obtained from the analysis of the magnetic force exerted by one current on another. This was used to define the practical unit of charge (1 coulomb) that is the foundation of electrical measurements in the MKS system.[1] In this scheme, with an appropriately defined unit of magnetic field, the basic law of force on a *moving* charge can then be written

$$\mathbf{F}_{\text{mag}} = q_2\mathbf{u} \times \mathbf{B} \qquad \text{(MKS system)} \qquad (8\text{–}3)$$

[1]The essential difference, of course, lies not in whether we use centimeters and grams rather than meters and kilograms, but in whether we base our definition of the unit charge on Coulomb's law or on the force between currents.

Given the value of **B** at a point, we can then calculate the magnetic force on any charged particle, with any velocity, at that point.

The *total* electromagnetic force on a charge at a given point can then be written as the sum of the electric force that it would experience if stationary, plus the magnetic force as given by Eq. (8–3). Thus, in general, we have

$$\mathbf{F} = q_2(\boldsymbol{\mathcal{E}} + \mathbf{u} \times \mathbf{B}) \qquad \text{(MKS system)} \tag{8–4}$$

a relation known as the *Lorentz force law*. We can regard Eq. (8–4) as a means of *defining* the electric field $\boldsymbol{\mathcal{E}}$ and the magnetic field **B** in terms of the forces found to be exerted on charges in various states of motion (including rest) at a given point.

The Coulomb force law is a complete statement of the force exerted on a charged particle, moving or stationary, by *stationary* charges. The magnetic force is associated with *moving source charges*. We see this explicitly in the interaction of current-bearing wires. A wire is electrically neutral (to an excellent approximation at least) whether or not it is carrying a current. It exerts no Coulomb force on a charged particle in its vicinity. But when it carries current, a magnetic field pattern is set up around it, and a force, as described by Eq. (8–3), is exerted on a moving charged particle nearby. The current is the systematic drift of electrons inside the wire, and we can ascribe the existence of the magnetic field to this motion. The force exerted by one current-bearing wire on another (both electrically neutral) can be understood, microscopically, in terms of the magnetic forces exerted on moving test charges in one wire by moving source charges in the other. Later in the chapter we shall develop this very result. As with the Coulomb force law, the designation of one charge as source charge and the other charge as test charge is of course arbitrary.

INTRODUCING THE RELATIVISTIC APPROACH

From the standpoint of any theory that regards all inertial frames as being physically equivalent, the distinction between effects due to stationary charges and effects due to moving charges cannot be accepted as fundamental. It is a distinction that depends on the use of a particular frame of reference. A test charge that is moving with respect to one frame, and therefore subject to a magnetic force, is stationary with respect to some other frame and therefore experiences no magnetic force even if a magnetic field exists at its location. Likewise, a source charge that moves with respect to one frame, and hence acts as the source of a magnetic field, is stationary with respect to another frame and therefore, as far as the latter frame is concerned, gives rise to an electric field only. It is always possible to choose a frame of reference in which any given particle is stationary. In the discussions that follow we shall repeatedly exploit this possibility, which allows us (in our imagination, at any rate) to convert any given problem into another one for which either the source charge or the test charge is stationary.

Our analysis will be carried out in terms of the *forces* that are exerted between charges in different states of motion. Our starting point (achieved, perhaps, by an initial shift of reference frames) will be the effect due to a stationary electric charge—i.e., an interaction described by the Coulomb force law and the associated electric field of the stationary charge. As observed from a different coordinate system, the "stationary" charge is moving. The transformation relations of relativity theory will enable us to discover how things are described in this new coordinate system. We shall learn that the law describing the electric field of a moving charge differs from that for a stationary charge. In addition, we shall see that a moving charge exerts a force upon a second moving charge that depends upon the velocities of *both* charges. This interaction will enable us to identify the *magnetic field* of the moving charge, whereas no such field exists for the stationary charge. It is in this sense that the correctness of Einstein's hunch, quoted at the head of this chapter, is demonstrated: What appears as a magnetic field in one coordinate system is nothing else but an electric field when viewed in some other coordinate system. *On the basis of Coulomb's law alone* (plus relativity) we can generate a quantitative description of the electric and magnetic interactions between charges moving with arbitrary constant velocities (and can also deal with many problems, although not all, involving accelerated charges).

Note especially, as you follow through the discussion, that it is the *forces* that are taken as primary. The identification of the electric and magnetic fields is then a secondary stage, based upon analysis of the total force into parts that are independent of the velocity of the test charge ($\boldsymbol{\mathcal{E}}$), and dependent on the velocity of the test charge ($\mathbf{B}$).

When we say that the force exerted by a stationary source charge upon a test charge is $\mathbf{F} = k(q_1 q_2/r^2)\mathbf{e}_r$, we need to be aware of the precise content of this statement. It describes the state of affairs at some given instant of time as measured in a certain reference frame—that frame in which q_1 is stationary. The vector $\mathbf{r}$ is the spatial separation of q_1 and q_2 at this instant, and the force $\mathbf{F}$ acts along the direction of $\mathbf{r}$. Under the action of this force alone, the momentum $\mathbf{p}_2$ of the test particle would change at a rate given by $d\mathbf{p}_2/dt = \mathbf{F}$; that is, $d\mathbf{p}_2/dt$ is a measure of the force $\mathbf{F}$. However, all the kinematic and dynamic quantities involved—the zero velocity of the source charge, the distance between the source charge and the test charge, and the time rate of change of the momentum of the test charge—are measured in a particular coordinate system. Referred to a second coordinate system which is moving relative to the first, the source charge will no longer be at rest, the distance (defined as being measured at a given instant in the new frame) between source and test charge may be different, and the rate of momentum change, the measure of the force on the test charge, may be different. If we can determine the new values of these various quantities as measured in the new coordinate system, then we will have a law which gives the force on a test charge in the presence of a moving source charge. The basis for all our calculations will be the Lorentz transformations, together with the force transformations developed in the last chapter. The procedure may seem awkward at first but is really not difficult.

We shall begin with some situations in which the test charge is stationary with respect to a reference frame S. Thus, as described in this frame, the only force that can act on it is due to an *electric* field. But let us imagine that the *source* charge moves with respect to S. Our procedure will be as follows:

1. Transform to a frame S' in which the source charge is stationary. Then the Coulomb law, Eq. (8–1), gives the force on the test charge as measured in S'. (The *test* charge is moving, as observed in S', but, as we have been careful to point out, the use of the Coulomb law requires only that the source charge should be stationary.)

2. Transform back from S' to S, so as to obtain a statement of the force, as measured in S, in terms of S coordinates only.

TABLE 8–1

$$\left.\begin{aligned} x' &= \gamma(x - vt) & x &= \gamma(x' + vt') \\ y' &= y & y &= y' \\ z' &= z & z &= z' \\ t' &= \gamma(t - vx/c^2) & t &= \gamma(t' + vx'/c^2) \end{aligned}\right\} \quad (8\text{–}5)$$

$$\left.\begin{aligned} u_x' &= \frac{u_x - v}{1 - vu_x/c^2} & u_x &= \frac{u_x' + v}{1 + vu_x'/c^2} \\ u_y' &= \frac{u_y/\gamma}{1 - vu_x/c^2} & u_y &= \frac{u_y'/\gamma}{1 + vu_x'/c^2} \\ u_z' &= \frac{u_z/\gamma}{1 - vu_x/c^2} & u_z &= \frac{u_z'/\gamma}{1 + vu_x'/c^2} \end{aligned}\right\} \quad (8\text{–}6)$$

$$\left.\begin{aligned} F_x' &= \frac{F_x - (v/c^2)(\mathbf{F} \cdot \mathbf{u})}{1 - vu_x/c^2} & F_x &= \frac{F_x' + (v/c^2)(\mathbf{F}' \cdot \mathbf{u}')}{1 + vu_x'/c^2} \\ F_y' &= \frac{F_y/\gamma}{1 - vu_x/c^2} & F_y &= \frac{F_y'/\gamma}{1 + vu_x'/c^2} \\ F_z' &= \frac{F_z/\gamma}{1 - vu_x/c^2} & F_z &= \frac{F_z'/\gamma}{1 + vu_x'/c^2} \end{aligned}\right\} \quad (8\text{–}7)$$

Our manipulations will make use of the transformation equations for position, time, velocity, and force. For convenience and repeated reference, therefore, we collect them in Table 8–1.

THE FORCE ON A STATIONARY TEST CHARGE

Case 1. Let the source charge q_1 be moving with the constant velocity $(v, 0, 0)$ relative to S and, at the time $t = 0$, be located instantaneously at the origin of S. Let the test charge q_2 be stationary on the x axis at the point $(x, 0, 0)$. This situation is shown in Fig. 8–2(a).

We cannot directly apply Coulomb's law to get the force on q_2, because q_1 is moving and the validity of Coulomb's law cannot be assumed when the source charge is in motion. In fact we shall see that Coulomb's law, applied directly, does *not* give the correct value for the force on the test charge.

However, let us look at the situation as it appears in S'. System S' is moving with velocity $\mathbf{v} = (v, 0, 0)$ relative to S. Therefore in S' the source charge q_1 is at rest at the origin, and the test charge q_2 is moving along the x' axis with velocity $\mathbf{u}' = (-v, 0, 0)$. At time t', the test charge will have space coordinates

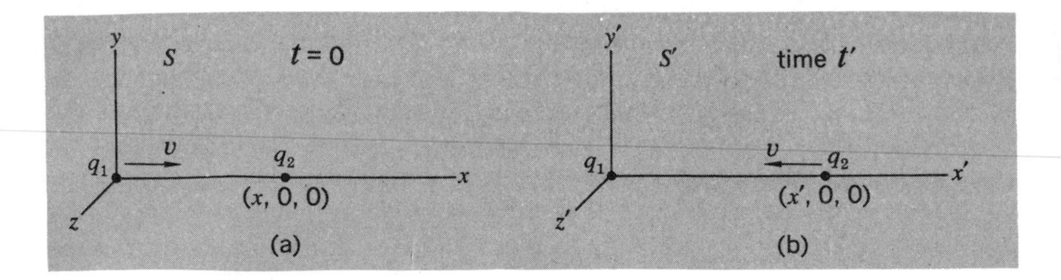

Fig. 8–2 Situation with moving source charge and stationary test charge (a) is transformed to another frame (b) in which the source charge is stationary and hence Coulomb's law can be applied.

$(x', 0, 0)$. We choose x' and t' so that $(x, 0, 0, 0)$ and $(x', 0, 0, t')$ describe the same space-time point relative to S and S', respectively. This situation is shown in Fig. 8–2(b).

Before going any further, let us draw attention to a very important feature, which could easily be overlooked. The aim of our calculation is to find the force exerted by q_1 on q_2 at a particular instant, $t = 0$, measured in S. If we describe a point event in terms of its coordinates (x, y, z, t), then the space-time coordinates of q_1 and q_2 are $(0, 0, 0, 0)$ and $(x, 0, 0, 0)$ respectively. We transform to S', using the Lorentz transformations:

$$x' = \gamma(x - vt) \qquad y' = y \qquad z' = z \qquad t' = \gamma(t - vx/c^2)$$

This gives the following results:

Space-time coordinates of q_1: $(0, 0, 0, 0)$

Space-time coordinates of q_2: $(x', 0, 0, t')$

where $x' = \gamma x$, $t' = -\gamma(vx/c^2)$. *The above coordinates do not define point events occurring at the same time in S'.* But to use the Coulomb force law in S', we must be able to specify the relative positions of q_1 and q_2 *at the same time* in that frame. We can do this because, in S', q_1 is stationary and hence its position at $t' = -\gamma vx/c^2$ is the same as its position at $t' = 0$. But an important matter of principle is involved here. Given two point events $(\mathbf{r}_1, t)$ and $(\mathbf{r}_2, t)$ occurring at the same time t in S, we can describe them as occurring at $(\mathbf{r}_1', t_1')$ and $(\mathbf{r}_2', t_2')$ in S', and in general $t_1' \neq t_2'$. In general, therefore, the value of $\mathbf{r}_2' - \mathbf{r}_1'$ does *not* represent a distance between two particles according to the criterion (simultaneity) for making such measurements in S'. Thus it is not, in general, the appropriate quantity to use in Coulomb's law—although it happens to be in the case we are discussing.

Let us return now to our problem, having justified the use of the distance $x' = \gamma x$ to describe the separation of q_1 and q_2 in S'. The force exerted on q_2 by q_1 is then described (in S') by the following components, according to Coulomb's law:

$$F_x' = kq_1q_2/x'^2 \qquad F_y' = 0 \qquad F_z' = 0$$

We proceed to transform back to S, using the equations in Table 8–1. For q_2 we have $u_x' = -v$. Substituting this, and the components of $\mathbf{F}'$ as given above, we find

$$F_x = F_x' = kq_1q_2/x'^2 \qquad F_y = 0 \qquad F_z = 0$$

In this case, therefore, the force on q_2 as measured in S at time $t = 0$ has the same value as in S' at time t'. However, the distance between q_1 and q_2, as measured in S at $t = 0$, is less than it is as measured in S' at time t'. This means that the *law of force* is modified if the source charge is moving. To see this modification, we must express F_x in terms of the separation of the charges as measured in S. Since $x' = \gamma x$,

$$F_x = \frac{1}{\gamma^2}\frac{kq_1q_2}{x^2}$$

Thus the source charge, moving with velocity v, acts on the test charge, at rest in the line of motion of the source charge, with a force $1/\gamma^2$ times the usual Coulomb force. This result holds good whether the source charge is moving toward or receding from the test charge.

Case 2. Consider now a second situation, in which the test charge q_2 is stationary on the y axis at the point $(0, y, 0)$. Again, let the source charge q_1 be moving with the constant velocity $(v, 0, 0)$ relative to S and, at the time $t = 0$, be located at the origin [Fig. 8–3(a)].

In S' the space-time coordinates of q_2, corresponding to the space-time point $(0, y, 0, 0)$ in S, are $(0, y', 0, 0)$, and the situation at $t' = 0$ is as shown in Fig. 8–3(b), with the velocity of q_2 given by $u_x' = -v$.

In S', q_1 is at rest, and therefore we can apply Coulomb's law to obtain the force on q_2 as observed in S', at the time $t' = 0$,

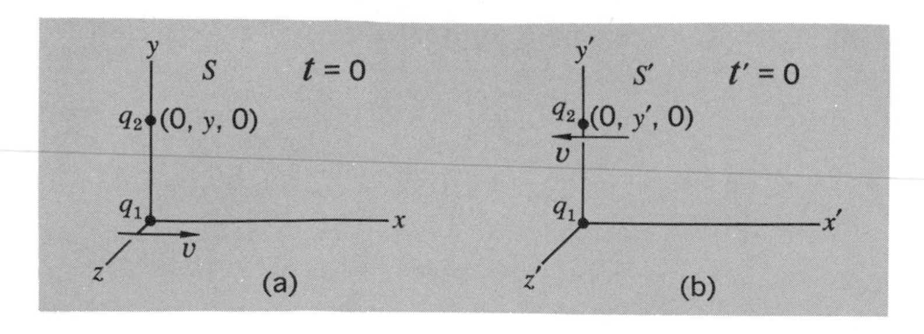

Fig. 8–3 Situation like that in Fig. 8–2 but with different geometry.

when q_2 is located at $(0, y', 0)$. This force is

$$F_x' = 0 \qquad F_y' = kq_1q_2/y'^2 \qquad F_z' = 0$$

Equations (8–7) now give us, for the force as measured in S,

$$F_x = 0 \qquad F_z = 0$$

$$F_y = \frac{F_y'/\gamma}{1 + vu_x'/c^2} = \frac{F_y'}{\gamma(1 - v^2/c^2)} = \gamma F_y'$$

In *this* case, therefore, the force on q_2 is different as measured in the two systems. But, since $y = y'$, the distance between the two charges, as measured in S and S' at $t = 0$ and $t' = 0$, is the same. By contrast, in case 1 the force as measured in the two systems remained the same, but the distance between the charges was different in the two systems. Either way, the force *law* is altered—in case 1 by a factor of $1/\gamma^2$, and in the present case, as we now see, by a factor of γ.

Expressing F_y in terms of the separation of the charges as measured in S, we have

$$F_y = \gamma \frac{kq_1q_2}{y^2}$$

Thus, at the instant when the line from the moving source charge to the stationary test charge is perpendicular to the direction of motion of the source charge, the force on the test charge is greater, by the factor γ, than the usual Coulomb force.

Case 3. In cases 1 and 2, the test charge had a special position relative to the source charge. Now let q_2 be stationary and located at any general point (x, y, z). Let q_1 be moving as before. Then in S, at $t = 0$, we have the situation shown in Fig. 8–4(a).

The space-time coordinates of q_2 are $(x, y, z, 0)$. In S', the same space-time point will have the coordinates (x', y', z', t') given by the coordinate transformations of Eqs. (8–5). In S' at time t', the picture will be as shown in Fig. 8–4(b).

In S', q_1 is stationary, and using Coulomb's law we have

$$F_x' = kq_1q_2\frac{x'}{r'^3} \qquad F_y' = kq_1q_2\frac{y'}{r'^3} \qquad F_z' = kq_1q_2\frac{z'}{r'^3}$$

From equations (8–7),

$$F_x = F_x' \qquad F_y = \gamma F_y' \qquad F_z = \gamma F_z'$$

Thus

$$F_x = kq_1q_2\frac{x'}{r'^3} \qquad F_y = \gamma kq_1q_2\frac{y'}{r'^3} \qquad F_z = \gamma kq_1q_2\frac{z'}{r'^3}$$

But

$$x' = \gamma x \qquad y' = y \qquad z' = z$$

Therefore, in S,

$$F_x = \gamma kq_1q_2x/r'^3 \qquad F_y = \gamma kq_1q_2y/r'^3 \qquad F_z = \gamma kq_1q_2z/r'^3$$

where

$$r' = (x'^2 + y'^2 + z'^2)^{1/2} = (\gamma^2x^2 + y^2 + z^2)^{1/2}$$

These results can be combined into a single vector equation:

$$\mathbf{F} = \gamma kq_1q_2\mathbf{r}/r'^3 = kq_1q_2\frac{\gamma\mathbf{r}}{(\gamma^2x^2 + y^2 + z^2)^{3/2}} \tag{8–8}$$

Thus the force on the stationary test charge q_2, at a given instant, due to the moving source charge q_1, is directed along the line to

Fig. 8–4 (a) Moving source charge and stationary test charge in arbitrary relative positions. (b) The same situation transformed to a frame in which the source charge is stationary.

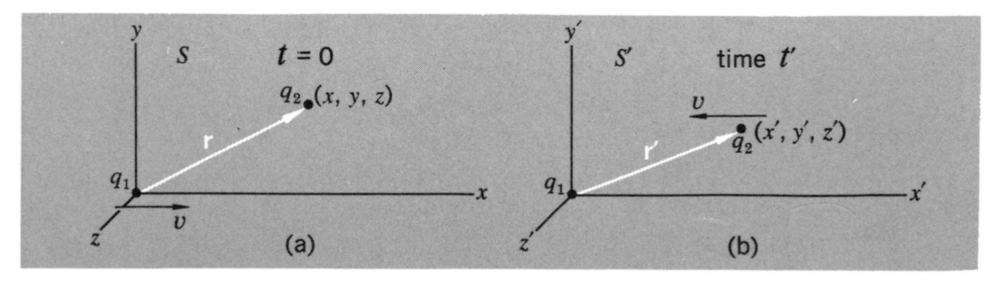

q_2 drawn from the position of q_1 at that same instant. As compared with the usual Coulomb force which would act on q_2 if q_1 were also stationary, the force in the present case is decreased by a factor of $1/\gamma^2$ if the test charge is in line ahead or behind the moving source charge, is increased by a factor of γ if q_2 is to the side of q_1, and varies continuously from the one value to the other for intermediate positions of q_2. It is interesting to note that the distortion of the field, as compared to that of a stationary charge, is symmetrical forward and backward—i.e., with respect to a plane passing through the charge at 90° to its direction of motion. One might intuitively have expected the field to be intensified forward and weakened backward, but this is not so.[1]

Since Eq. (8–8) describes the force exerted on a *stationary* test charge, the value of $\mathbf{F}/q_2$ is just the electric field $\boldsymbol{\mathcal{E}}$ due to q_1. Thus we have

$$\boldsymbol{\mathcal{E}}(x, y, z) = \gamma k q_1 \mathbf{r}/r'^3 = k q_1 \frac{\gamma \mathbf{r}}{(\gamma^2 x^2 + y^2 + z^2)^{3/2}} \tag{8–9}$$

Now the electric field due to a stationary source charge is radial and, of course, spherically symmetrical; that is, it is the same in all directions. It is simply the Coulomb field $\boldsymbol{\mathcal{E}} = k q_1 \mathbf{r}/r^3$. If the source charge is moving uniformly, the electric field is no longer spherically symmetrical. Its strength is different in different directions. But, at each instant, the direction of the electric field is still radial with respect to the position of the source charge at that same instant.

If you think about this last result a bit—that at each instant the electric field due to a uniformly moving source charge is directed radially away from the position of the source charge *at that same instant*—you may begin to realize that this is a very surprising result.

Consider what is being said. A charge q_1 is moving with constant speed along the x axis (Fig. 8–5). Suppose that, at the

[1]See, however, Problem 8–3.

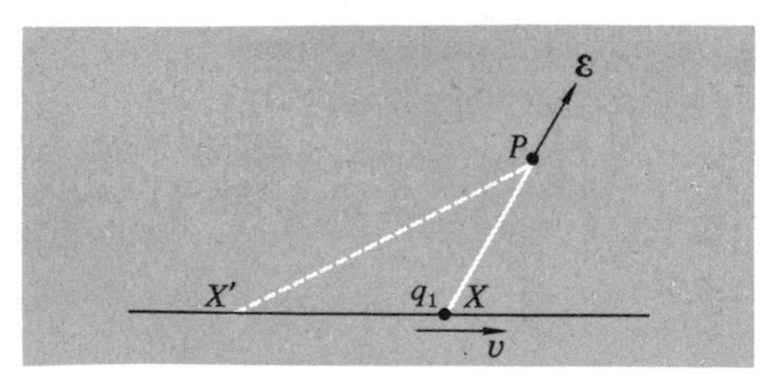

Fig. 8–5 Electric field at P, ascribable to a uniformly moving charge q_1 at the time when it was at the position X′, is directed radially from the present position, X, of the charge.

present instant, q_1 is located at point X. The electric field $\mathcal{E}$ at a point P at the present instant will be directed radially away from X as shown. That means if a test charge q_2 were placed at P, it would feel a force $\mathcal{E}q_2$ acting on it in the direction away from X. That all *seems* very reasonable.

But if we believe that no effect—no mass, no energy, no force—can be transmitted with a speed greater than c, then the test charge q_2, located at P, cannot "know" the present position of q_1. Some time must elapse before any signal, any effect, originating from the location and behavior of q_1 at its present position X will arrive at P—and by then q_1 will no longer be at X.

Let X' be an earlier position of q_1—a position such that a signal, leaving q_1 at the instant it was located at X', and then traveling with the speed c, arrives at P at the present instant. And at the present instant q_1, continuing its motion with speed v, arrives at X. Both trips take the same time. This means that

$$\frac{X'P}{c} = \frac{X'X}{v} \qquad \text{or} \qquad \frac{X'X}{X'P} = \frac{v}{c}$$

X' is called the *retarded position* of q_1 with respect to point P. Any signal which left q_1 at some point in its journey between X' and X will not have reached P by the present instant. (Prove that.) The field, *now*, at P must have originated from the behavior of q_1 at position X' or perhaps if the signal velocity were less than c, from its behavior at an earlier position. But nothing the charge q_1 may have done after the moment it was at X' can have any effect at P at the present instant.

Nevertheless, the field at P points away from the *present position* of q_1. Nature behaves in such a way that, for a uniformly moving source charge, even though the field produced at some point P originated from the location and behavior of the source charge at an *earlier time*, nevertheless the field points away from the position of the source charge at the present time. It is as though nature calculates where the source charge should be at the present time and acts accordingly. (Indeed, it may already have occurred to you that, unless the charge does continue to move with constant velocity after passing X', it may not in fact *be* at X at the instant when the field at P is observed. Nevertheless, X will still be the point from which the field appears to originate at this instant.) Thus a result which at first glance may seem rather obvious is seen, upon closer examination, to be quite surprising—but nevertheless true.

THE FORCE ON A MOVING TEST CHARGE

Coulomb's law tells us the force on a test charge, stationary or moving, due to a *stationary* source charge. In the preceding calculations, our procedure has been, in effect, to set up just such a situation in a system S'—a stationary source charge and a moving test charge. The effect on the test charge was given by Coulomb's law. We then transformed to a different inertial system S. But in each case, we restricted our choice of S so that the *test charge* was at rest in S. In this manner we were able to find the force on a stationary test charge due to a uniformly moving source charge.

Now we shall investigate the effect of a *uniformly moving source charge* upon a *moving test charge*. Again our starting point, as far as the calculation of forces is concerned, will be a system S' in which the source charge is at rest, so that Coulomb's law gives a complete and accurate specification of the force. But then we shall transform back to a system S in which both the source charge and the test charge are in motion. We shall find a new feature: The force on the test charge includes a term dependent on the velocity of the test charge. This is exactly the situation when a charge moves in a magnetic field, and we shall use it to identify and evaluate the magnetic field.

We shall start with some simple cases so that we may unveil the properties of this new feature without becoming involved in any extended algebraic manipulations.

Case 1. Let the source charge q_1 and test charge q_2 both be moving with the same constant velocity $(v, 0, 0)$ relative to S. At the time $t = 0$ let the line connecting the instantaneous positions of q_1 and q_2 coincide with the y axis. This situation is shown in Fig. 8–6(a). In S' the situation is extremely simple, because both charges are stationary and their separation is

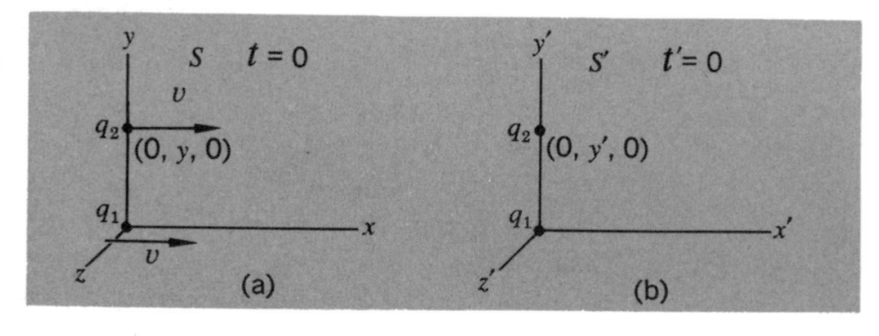

Fig. 8–6 (a) Two charges move side by side with equal velocities. (b) The situation is transformed to the frame in which both charges are at rest.

$y' = y$ [Fig. 8–6(b)]. In S', we have

$$F_x' = 0 \qquad F_y' = kq_1q_2/y'^2 \qquad F_z' = 0$$

Since, for q_2, we have $\mathbf{u}' = 0$, the force transformations, equations (8–7), give us

$$F_x = 0 \qquad F_y = F_y'/\gamma \qquad F_z = 0$$

from which we find

$$F_y = \frac{1}{\gamma}\frac{kq_1q_2}{y^2}$$

Now with q_2 *stationary*, we should have

$$F_y = \gamma\frac{kq_1q_2}{y^2}$$

as calculated in case 2 of the previous section [or by reference to Eq. (8–8) for the general case of a stationary test charge]. The difference between these two values of F_y represents a *magnetic force* exerted on the moving test charge by the moving source charge:

$$F_{\text{mag}} = \frac{kq_1q_2}{y^2}\left(\frac{1}{\gamma} - \gamma\right) = -\frac{v^2}{c^2}\frac{\gamma kq_1q_2}{y^2}$$

The negative sign shows that, if q_1 and q_2 are of like sign, the magnetic force is one of attraction between charges traveling side by side in the same direction. Although one should not read too much into such a special case as we have chosen, it is suggestive that this result can be written

$$F_{\text{mag}} = -\frac{v^2}{c^2}F_{\text{elec}}$$

where F_{elec} ($= \gamma kq_1q_2/y^2$) is the force that would be exerted on q_2 if it were stationary. Is the size of the magnetic force always proportional to the size of the electric force, and what is the origin of the factor v^2/c^2? The answers to both these questions will emerge as we consider other cases.

Case 2. Consider now a problem identical with case 1 except for one thing—the charge q_2 has a velocity of magnitude u_x, different from v but still parallel to it.[1] The situations as ob-

[1]Why not consider a case in which q_2 is on the x axis? Analyze such a case for yourself and you will discover why it is not very instructive.

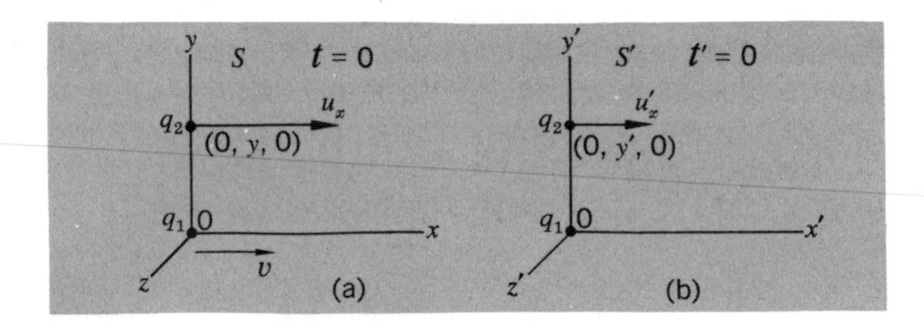

Fig. 8–7 (a) Two charges move side by side with different velocities in the same direction. (b) The situation is transformed to a frame in which the source charge is stationary.

served in S and S' are then as shown in Fig. 8–7. In S', we have, once again,

$$F_x' = 0 \qquad F_y' = kq_1q_2/y'^2 \qquad F_z' = 0$$

but in transforming back to S by equations (8–7) we must make use of the nonzero value of $\mathbf{u}'$:

$$u_x' = \frac{u_x - v}{1 - vu_x/c^2} \qquad (u_y' = 0, u_z' = 0)$$

Using these, we find

$$F_x = 0 \qquad F_y = \frac{\gamma k q_1 q_2}{y^2}\left(1 - \frac{vu_x}{c^2}\right) \qquad F_z = 0$$

This is a more revealing result than case 1. It shows us that the magnetic force exerted on q_2, if it lies on the y axis and is traveling parallel to x, is given by

$$F_{\text{mag}} = -\frac{vu_x}{c^2} F_{\text{elec}}$$

The magnetic force is seen explicitly to be proportional to the speed of the test charge, just as the force law, Eq. (8–3), would require. Also the factor vu_x/c^2, which expresses the magnetic force in this case as a fraction of the electric force, is proportional to the speed of the source charge, too.

Case 3. We shall consider yet another problem quite similar to the last two. Again we shall place q_2 at the point $(0, y, 0)$ in

S, but this time we shall assume that its motion is along the y axis: $\mathbf{u} = (0, u_y, 0)$. The situations as observed in S and S' are thus as shown in Fig. 8–8. Once more, we have

$$F_x' = 0 \qquad F_y' = kq_1q_2/y'^2 \qquad F_z' = 0$$

but now there are *two* nonzero-velocity components of q_2 as measured in S':

$$u_x' = -v \qquad u_y' = u_y/\gamma \qquad u_z' = 0$$

Using these, we find

$$F_x = \frac{vu_y}{c^2}\frac{\gamma kq_1q_2}{y^2} \qquad F_y = \frac{\gamma kq_1q_2}{y^2} \qquad F_z = 0$$

The situation is now really beginning to open up. Looking at the results of this problem and the previous one, we see that the total force on q_2 can be regarded as made up of two parts:

1. An electric force, along the line from q_1 to q_2, of magnitude $\gamma kq_1q_2/r^2$.

2. A magnetic force, at right angles to the direction of motion of q_2, proportional to the electric force and proportional also to the magnitude of the velocity of q_2 (as well as to the speed v of q_1).

Of course, we are still a long way from considering a truly general case, with q_2 at an arbitrary position with arbitrary velocity. But the next case will reinforce the conclusions that we have just proposed.

Fig. 8–8 (a) A source charge and a test charge have velocities of different magnitudes at right angles to one another. (b) The situation is transformed to a frame in which the source charge is stationary.

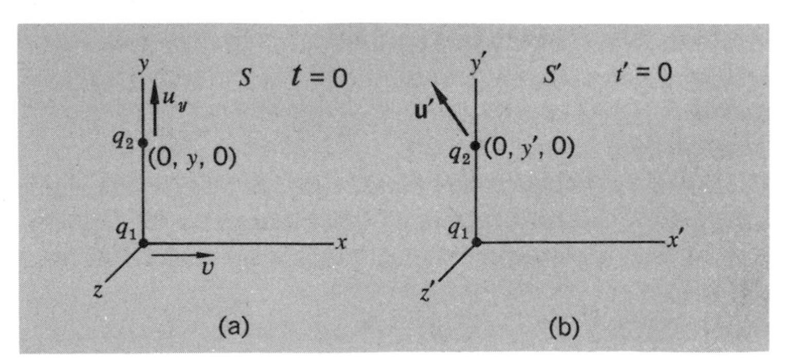

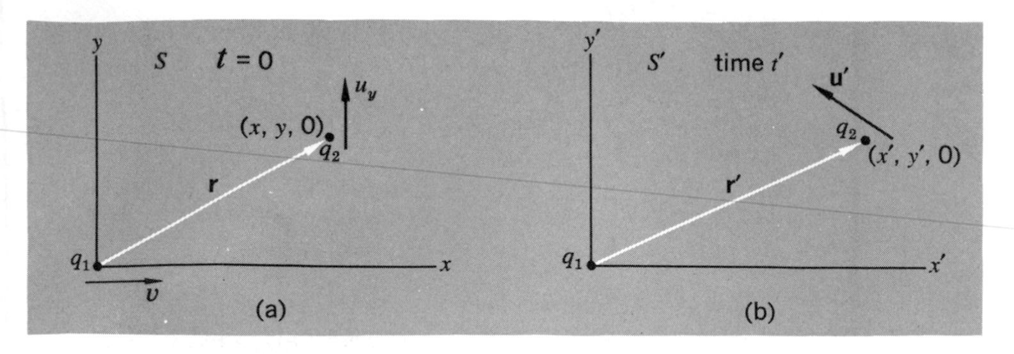

Fig. 8–9 Situation like that in Fig. 8–8, but the relative positions of the charges are now arbitrary.

Case 4. Let q_1, as usual, be at the origin of S at $t = 0$, with velocity $\mathbf{v}$ along x, i.e., $v = (v, 0, 0)$. At this same time let q_2 be at the point $(x, y, 0)$ and moving parallel to the y axis: $\mathbf{u} = (0, u_y, 0)$. The situations as observed in S and S' are shown in Fig. 8–9. From the Coulomb force law, as applied in S', we have

$$F_x' = kq_1q_2x'/r'^3 \qquad F_y' = kq_1q_2y'/r'^3 \qquad F_z' = 0$$

The velocity $\mathbf{u}'$ is the same as in the previous case:

$$u_x' = -v \qquad u_y' = u_y/\gamma \qquad u_z' = 0$$

Use of the force transformations, equations (8–7), then leads to the following results, as may be confirmed without too much algebra:

$$F_x = \frac{\gamma kq_1q_2}{r'^3}\left(x + \frac{vu_y}{c^2}y\right) \qquad F_y = \frac{\gamma kq_1q_2}{r'^3}y \qquad F_z = 0$$

where $r' = (\gamma^2x^2 + y^2)^{1/2}$.

Let us look carefully at this result. In it we can recognize first the electric force $\mathbf{F}_{\text{elec}}$, given by

$$\mathbf{F}_{\text{elec}} = \frac{\gamma kq_1q_2\mathbf{r}}{(\gamma^2x^2 + y^2)^{3/2}}$$

exactly according to Eq. (8–8). The remaining part is a force in the x direction—at right angles to the velocity of q_2, once more, and proportional to the magnitude of this velocity:

$$F_{\text{mag}} = \frac{\gamma kq_1q_2}{r'^3}\frac{vu_y}{c^2}y$$

Can we relate it simply, as before, to the electric force? Yes, we can. And this less special geometry reveals more of the essentially vectorial relationship involved. Let us follow the lead suggested by what we already know about the magnetic force, as expressed in Eq. (8–3):

$$\mathbf{F}_{\text{mag}} = q_2\mathbf{u} \times \mathbf{B}$$

In the present case, the velocity $\mathbf{u}$ is parallel to y, and the magnetic force (as always, perpendicular to it) is parallel to x. This would require the vector $\mathbf{B}$ to be parallel to z. Moreover, its magnitude is given by

$$B = \frac{\gamma k q_1}{r'^3}\frac{vy}{c^2}$$

Now the product vy has exactly the magnitude of the vector product $\mathbf{v} \times \mathbf{r}$, where $\mathbf{v}$ is the velocity of q_1. Moreover, the vector $\mathbf{v} \times \mathbf{r}$ *is* along the z direction. Thus we can provide a correct description of $\mathbf{B}$, in both magnitude and direction, through the equation

$$\mathbf{B} = \frac{1}{c^2}\mathbf{v} \times \left(\frac{\gamma k q_1}{r'^3}\mathbf{r}\right)$$

But the quantity in parentheses is none other than the electric field $\boldsymbol{\mathcal{E}}$ due to the moving charge q_1, as given by Eq. (8–9). Hence we can write

$$\mathbf{B} = \frac{1}{c^2}(\mathbf{v} \times \boldsymbol{\mathcal{E}}) \qquad (8\text{–}10)$$

where

$$\boldsymbol{\mathcal{E}} = \gamma k q_1 \mathbf{r}/r'^3 \qquad (r'^2 = \gamma^2 x^2 + y^2)$$

and then the total force on the moving test charge is given by

$$\mathbf{F} = q_2\boldsymbol{\mathcal{E}} + q_2\mathbf{u} \times \mathbf{B}$$

precisely as in Eq. (8–4), with $\boldsymbol{\mathcal{E}}$ and $\mathbf{B}$ as given by Eqs. (8–9) and (8–10).

It can be objected, of course, that we have not really *proved* these results. But an analysis of the case in which q_2 has any position and any velocity will be found to accord once more with our equations. The combination of electric and magnetic forces on a moving charge, caused by another moving charge, can be exactly calculated from a knowledge of the Coulomb force in a frame in which the source charge is at rest.

In the remainder of this chapter we shall apply these basic results to the problem of extended currents—i.e., long lines of moving charges—so as to see how, on the basis of relativity, one can obtain a very illuminating picture of the mechanism by which one current-carrying wire exerts a force upon another. We shall approach this by first considering the properties of a long line of charges all of the same sign.

THE FIELD OF A LINE DISTRIBUTION OF CHARGE

We shall consider the field due to a uniform distribution of charge along the x axis. In reality, it is impossible to have exactly this kind of distribution, because we know that electric charge is quantized in multiples of e. However, since e is so small by macroscopic standards, this approximation will be justifiable for many purposes. We shall first consider a stationary set of charges, and then see what modifications arise if the whole line of charge is moving with some constant velocity.

Stationary line of charge

We shall assume a constant linear density of charge λ (measured, for example, in coulombs per meter of distance) extending along the x axis from $-\infty$ to $+\infty$ (Fig. 8–10). We shall ask: What is the electric field at a point P located a distance b away from the x axis?

By the principle of superposition, the field at P is the vector sum of the fields produced by every element of charge along the line. Clearly the resultant x component of the field at P is zero, from symmetry considerations. But every element of charge contributes a positive y component of field at P.

The amount of charge in the line element dx is $\lambda\, dx$. The

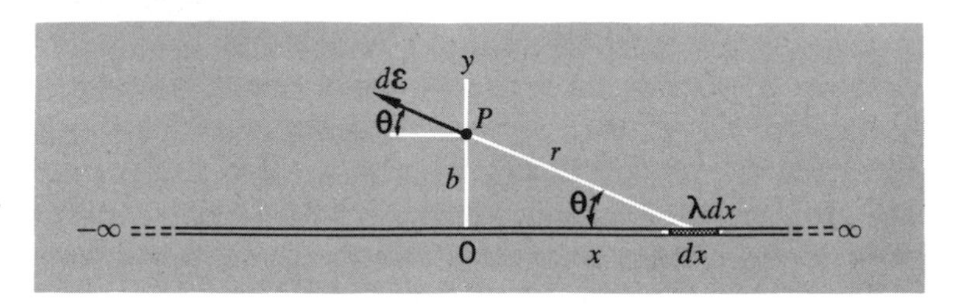

Fig. 8–10 Calculation of electric field due to an infinite line of stationary charge.

field it produces at P is

$$d\boldsymbol{\mathcal{E}} = k \frac{\lambda\, dx}{r^2} \mathbf{e}_r$$

and the y component is

$$d\mathcal{E}_y = k \frac{\lambda \sin\theta\, dx}{r^2}$$

Let us express this in terms of the one variable θ:

$$r = b/\sin\theta$$

$$x = b\cot\theta$$

$$dx = -b\csc^2\theta\, d\theta$$

and therefore

$$d\mathcal{E}_y = -\frac{k\lambda}{b}\sin\theta\, d\theta$$

The resultant y component of field is then

$$\mathcal{E}_y = -\frac{k\lambda}{b}\int_{\pi}^{0}\sin\theta\, d\theta = \frac{2k\lambda}{b}$$

Since this is the only nonvanishing field component, we have, for the complete line of charge,

$$\boldsymbol{\mathcal{E}} = 2k\frac{\lambda}{b}\mathbf{e}_y \tag{8-11}$$

Line of moving charges

The only field we needed to consider above was the electric field. There is no magnetic field because the charges are all at rest. But what happens to the electric field and what is the magnetic field if the charges are moving?

Again, let us consider a constant linear charge density λ extending along the x axis from $-\infty$ to $+\infty$, but this time we shall let the entire charge distribution move along the x axis with the velocity v [Fig. 8–11(a)].

As we have seen, moving charges produce electric fields which differ from the normal Coulomb fields, and in addition they produce magnetic fields. The moving charge density in the vicinity of point 1 produces an electric field $d\boldsymbol{\mathcal{E}}_1$ at P which is larger than the normal Coulomb field by a factor of γ. The mov-

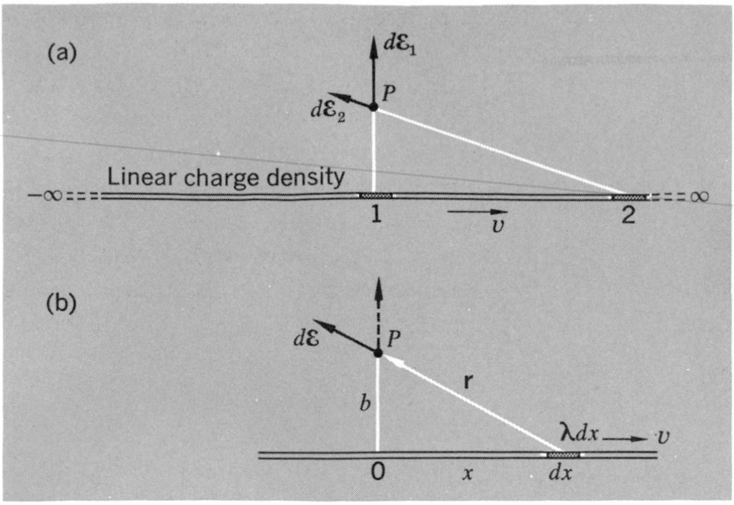

Fig. 8–11 *(a) Electric fields due to near and distant parts of an infinite line of moving charge. (b) Electric field of a line element at an arbitrary value of x.*

ing charge density far away along the x axis (as at point 2) produces an electric field $d\mathcal{E}_2$ at P which is less than the normal Coulomb field by a factor approaching $1/\gamma^2$. [If you want to remind yourself of these results, refer to Eq. (8–9).] It is not immediately obvious which of these two opposing effects is dominant. Again, let us integrate the differential contributions to obtain the resultant field.

The contribution to the electric field at P due to the quantity of charge $\lambda\, dx$ moving with velocity v at some arbitrary point along the line [Fig. 8–11(b)] is given by Eq. (8–9) and is as follows:

$$d\boldsymbol{\mathcal{E}} = k(\lambda\, dx)\frac{\gamma \mathbf{r}}{(\gamma^2 x^2 + b^2)^{3/2}}$$

As with the line of stationary charges, when we add all the contributions to the x component of field from $x = -\infty$ to $x = +\infty$, then $\mathcal{E}_x$ will vanish. It is worth noting that this is so in spite of the fact that the velocity points in a particular direction along x. This is an expression of the forward-backward symmetry of the electric field of a moving charge. As for $\mathcal{E}_y$, we have

$$\mathcal{E}_y = \int d\mathcal{E}_y = \int_{-\infty}^{+\infty} \frac{k\gamma\lambda b\, dx}{(\gamma^2 x^2 + b^2)^{3/2}}$$

which gives

$$\mathcal{E}_y = 2k\frac{\lambda}{b} \qquad (8\text{–}12)$$

The electric field of this moving array of charge is exactly the same as that of the stationary array! The two opposing effects—the factor of γ from the nearby charges and the factor of $1/\gamma^2$ from the distant charges—compensate each other exactly.

Now how about the magnetic field? The stationary array, of course, produces no magnetic field. Remember what this means—simply that the stationary charges exert exactly the same force on moving charges as on stationary ones. But the moving linear array does produce a magnetic field. We have seen [Eq. (8–10)] that the magnetic field of a uniformly moving charge is

$$\mathbf{B} = \frac{1}{c^2}\mathbf{v} \times \boldsymbol{\mathcal{E}}$$

where $\boldsymbol{\mathcal{E}}$ is the electric field produced by the charge and $\mathbf{v}$ is the constant velocity of the charge. Thus, if the element of charge $\lambda\, dx$ produces the small electric field $d\boldsymbol{\mathcal{E}}$, then its contribution to the magnetic field is

$$d\mathbf{B} = \frac{1}{c^2}\mathbf{v} \times d\boldsymbol{\mathcal{E}}$$

The magnetic field of the entire linear array is then

$$\mathbf{B} = \int d\mathbf{B} = \int \frac{1}{c^2}\mathbf{v} \times d\boldsymbol{\mathcal{E}}$$

But $\mathbf{v}$ is a constant velocity along the x axis and is the same for all the elements of charge along the line. Thus we can put

$$\mathbf{B} = \frac{1}{c^2}\mathbf{v} \times \int d\boldsymbol{\mathcal{E}}$$

and therefore

$$\mathbf{B} = \frac{1}{c^2}(\mathbf{v} \times \boldsymbol{\mathcal{E}})$$

where $\boldsymbol{\mathcal{E}}$ is the field produced by the entire linear array, as given by Eq. (8–12). Since $\boldsymbol{\mathcal{E}}$ is everywhere radially outward from the x axis and $\mathbf{B}$ is perpendicular both to $\boldsymbol{\mathcal{E}}$ and to $\mathbf{v}$, the magnetic field lines consist of concentric circles around the x axis—corresponding, of course, to one of the familiar facts of experimental electromagnetism.

In the next section we shall consider the magnetic *force* exerted by such a moving charge array on a single moving charge traveling parallel to the line of charge at a velocity $\mathbf{u}$. We can anticipate the actual result by taking the above equation for **B** together with the general magnetic force law, Eq. (8–3). This gives us

$$\mathbf{F}_{\text{mag}} = -q_2 \frac{uv}{c^2} \mathbf{\mathcal{E}} = -q_2 \frac{uv}{c^2} \left(2k \frac{\lambda}{b} \right) \mathbf{e}_y \tag{8–13}$$

if **u** and **v** are in the same direction. Thus if q_2 has the same sign as the line of charges, the magnetic force is attractive in this case.

THE MAGNETIC FIELD AND RELATIVITY

The preceding calculation for **B** and the associated magnetic force could hardly be simpler. Nevertheless we shall approach the problem in yet another way to emphasize the relativistic nature of the magnetic field.

To investigate the magnetic field at P due to the moving linear distribution of charge, we must place a *moving* test charge at P. To simplify matters, we shall let the test charge q_2 move *parallel to the x axis* with the velocity **u** [Fig. 8–12(a)].

In this reference frame S, the force on q_2 contains both an electric part and a magnetic part:

$$\mathbf{F} = \mathbf{F}_{\mathcal{E}} + \mathbf{F}_B \tag{8–14}$$

The electric force is, by Eq. (8–11), given by

$$\mathbf{F}_{\mathcal{E}} = \mathcal{E} q_2 = 2k \frac{\lambda}{b} q_2 \mathbf{e}_y \tag{8–15}$$

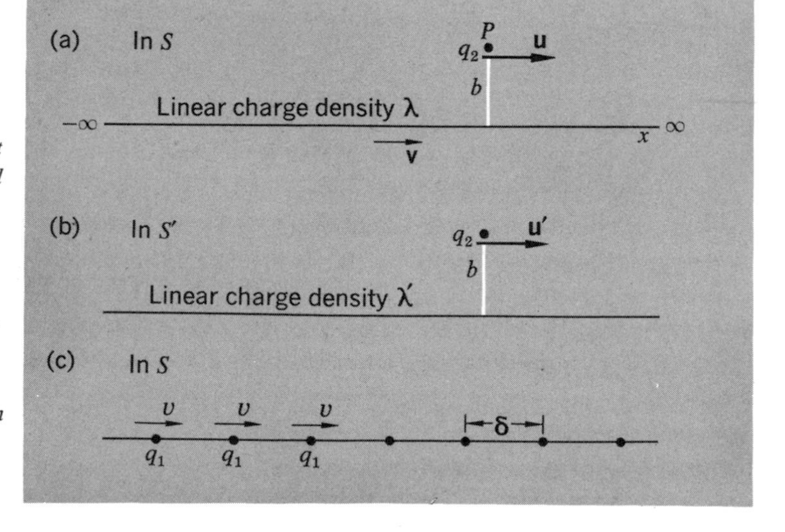

Fig. 8–12 (a) A test charge moves parallel to a line of moving charge. (b) The situation is transformed to a frame in which the line charge is stationary. (c) The linear charge density q_1/δ is not an invariant.

The magnetic force is

$$\mathbf{F}_B = q_2 \mathbf{u} \times \mathbf{B} \tag{8-16}$$

and if we can find $\mathbf{F}_B$, then we shall know something about $\mathbf{B}$.

Let us look at the system from a reference frame S' moving with velocity $\mathbf{v}$ relative to S. In S', the linear array is stationary and q_2 is moving with a velocity $\mathbf{u}'$ [Fig. 8–12(b)]. The distance from q_2 to the line is transverse to the direction of relative motion between S' and S and therefore is unchanged. But what happens to the linear charge density along the line? We can approximate the linear array by spacing an infinite collection of moving discrete charges q_1 a distance δ apart in frame S, and choosing q_1 and δ such that $q_1/\delta = \lambda$ [Fig. 8–12(c)].

If now we let q_1 and δ both approach zero in such a way that the ratio q_1/δ always remains constant and equal to λ, then we approach our idealized array—a line distribution of charge with constant linear density λ, moving with velocity $\mathbf{v}$.

In S' the charges q_1 are at rest. The distances between them are therefore greater, by the factor γ, than they are in S. Hence the charge density in S' is $\lambda' = q_1/\gamma\delta = \lambda/\gamma$ and is less than it is in S.

Now in S' the source charges are stationary. Thus the entire force on q_2 is purely electric and is given [cf. Eq. (8–11)] by

$$F' = 2k\frac{\lambda'}{b}q_2 = \frac{1}{\gamma}\left(2k\frac{\lambda}{b}\right)q_2$$

The force F' is in the y direction. Equations (8–7) tell us how to transform F' to find the total force F on q_2 as observed in S:

$$F_y' = \frac{F_y}{\gamma(1 - u_x v/c^2)}$$

Applied to the present case, this gives

$$F_y = \gamma\left(1 - \frac{uv}{c^2}\right)F_y'$$

That is,

$$\mathbf{F} = 2k\frac{\lambda}{b}q_2\left(1 - \frac{uv}{c^2}\right)\mathbf{e}_y \tag{8-17}$$

Thus from Eqs. (8–14), (8–15) and (8–17), the magnetic force observed in S is

$$\mathbf{F}_B = \mathbf{F} - \mathbf{F}_{\mathcal{E}}$$

Therefore,

$$\mathbf{F}_B = -\frac{uv}{c^2}\left(2k\frac{\lambda}{b}\right)q_2\mathbf{e}_y = -q_2\frac{uv}{c^2}\mathcal{E}$$

in complete agreement with Eq. (8–13). Using Eq. (8–16), we can then infer that the y component of $\mathbf{B}$ is zero, and that its z component is given by

$$B_z = \frac{v}{c^2}\left(2k\frac{\lambda}{b}\right) = \frac{1}{c^2}v\mathcal{E} \qquad (8\text{–}18)$$

in agreement with Eq. (8–10). To verify that this is the whole of $\mathbf{B}$—that the x component of $\mathbf{B}$ is zero—we should have to let our test charge q_2 move in some other direction, say the y direction. But we shall leave that as an exercise for the reader. Once again we see that a force which is of purely electric origin in one frame (S') has both electric and magnetic constituents from the standpoint of another frame (S).

THE MAGNETIC FORCE ON A MOVING CHARGE DUE TO A CURRENT-BEARING WIRE

For ordinary velocities, the magnetic force between two electric charges is very, very small compared to the electric force—smaller by the factor uv/c^2, for example, for charges moving side by side at speeds u and v. It can be easily observed only if we can manage to get rid of the electric force. Fortunately we can do that, for nature provides both positive and negative charges.

Consider, for example, a long, electrically neutral copper wire in which a current is flowing. The positive copper ions remain stationary and the free negative electrons move, say with a velocity $\mathbf{v}$.[1] What is the force on a moving test charge outside the wire? For simplicity let us first consider a test charge q_2 (e.g., an electron) moving at the same velocity $\mathbf{v}$ as we have assumed for the electrons in the wire [Fig. 8–13(a)]. The densities of positive and negative charges in the wire are equal and opposite, say $\pm\lambda_0$. They produce electric fields, at q_2, which are equal in magnitude ($\mathcal{E} = 2k\lambda_0/b$) but opposite in direction. Therefore the electric force on q_2 is zero. If q_2 were stationary, that would

[1]This is, of course, a gross oversimplification. The conduction electrons in a wire have all sorts of speeds and directions, but $\mathbf{v}$ represents a steady mean *drift* velocity associated with the net current flow.

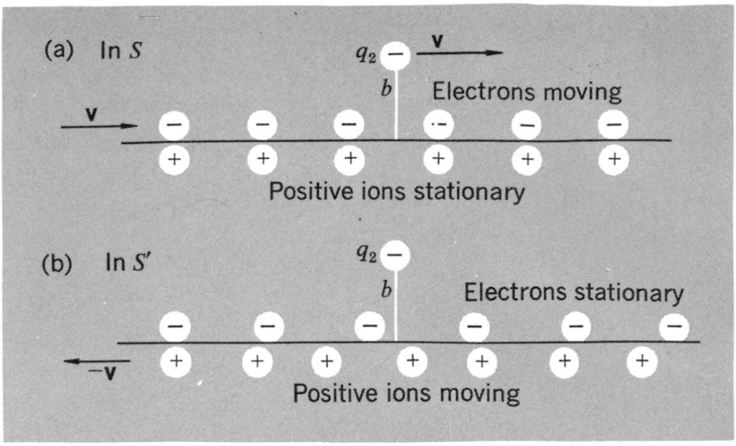

Fig. 8–13 (a) A test charge moves parallel to a current-bearing wire. (b) The situation is transformed to a frame in which the test charge is stationary.

be the end of it. But q_2 is moving and therefore can "feel" the magnetic field.

To find out something about the magnetic force on q_2 as observed in S, let us shift to a different frame, S', moving with velocity $\mathbf{v}$ with respect to S. In S', q_2 is stationary, the electrons in the wire are stationary, and the positive ions are moving with velocity $-\mathbf{v}$ [Fig. 8–13(b)].

As before, the distance between the stationary electrons in the wire is (from the standpoint of S') increased and therefore their linear density λ_- is decreased by the factor $1/\gamma$: $\lambda_- = -\lambda_0/\gamma$. On the other hand, the relativistic contraction of the distance between the moving positive ions results in an *increase* in their linear density λ_+ by the factor γ: $\lambda_+ = \gamma\lambda_0$. The positive and negative charge densities as observed in *this* frame are no longer equal and opposite! And the difference is in the direction that results in a net force on q_2 that is an attractive force if it is an electron traveling in the same direction as the electrons in the wire.

Let us now spell this out in detail. In S' the test-charge q_2 is stationary and the force on it, F', depends only on the electric field of the linear arrays of positive and negative charge. The net charge density λ' as measured in S' is given by

$$\lambda' = \lambda_+ + \lambda_- = \lambda_0\left(\gamma - \frac{1}{\gamma}\right)$$

Thus for the net force F' we have

$$F_y' = 2k\frac{\lambda'}{b}q_2 = \frac{2k\lambda_0}{b}\left(\gamma - \frac{1}{\gamma}\right)q_2$$

Transforming back to S, in the now familiar way, we find that the force on q_2, observed in S, is given by

$$F_y = \frac{F_y'}{\gamma} = \frac{2k\lambda_0}{b}\left(1 - \frac{1}{\gamma^2}\right) q_2$$

But

$$1 - \frac{1}{\gamma^2} = \frac{v^2}{c^2}$$

Therefore,

$$F_y = \frac{v^2}{c^2}\left(\frac{2k\lambda_0}{b}\right) q_2 = \beta^2 \mathcal{E} q_2,$$

where $\mathcal{E}$, remember, is the magnitude of the field, in S, of either the positive or the negative charges in the wire. Since there is no net electric force on q_2 in S, the force $F\ (= \beta^2 \mathcal{E} q_2)$ is entirely the magnetic force. It is down by a factor β^2 from the electric force which either the positive ions or the moving electrons alone would have exerted, but is observable even for small values of β because the electric force has been exactly canceled out by the presence of both positive and negative charges.

We see, in this example, that the magnetic field observed in S is related to the electric field observed in S'. And the electric field in S' arises from the relativistic changes in the distances between the moving charges—the Lorentz contraction effect.

This becomes a really astounding result when we consider it quantitatively. How fast are the electrons moving along a typical current-carrying wire? Consider a copper wire with a cross-sectional area of about 1 mm^2 ($= 10^{-2}$ cm^2) carrying a current of 10 amp. A current of 10 amp means that 10 coulombs/sec, or

$$\frac{10 \text{ coulombs/sec}}{1.6 \times 10^{-19} \text{ coulomb/electron}} \approx 6 \times 10^{19} \text{ electrons/sec}$$

must pass through any cross-sectional area of the wire. In solid copper there is about one free electron for every Cu atom, or about 10^{23} free electrons/cm^3. (Check this.) If the electrons are moving along the wire with a speed v then, in a time interval Δt, all the free electrons in the cylinder of base area 10^{-2} cm^2 and of length $v\,\Delta t$ (Fig. 8–14) will pass by position B. This number is $10^{23} \times 10^{-2} v\,\Delta t$. But the number of electrons that must pass B when a current of 10 amp flows for a time Δt is

$$6 \times 10^{19} \frac{\text{electrons}}{\text{sec}} \times \Delta t$$

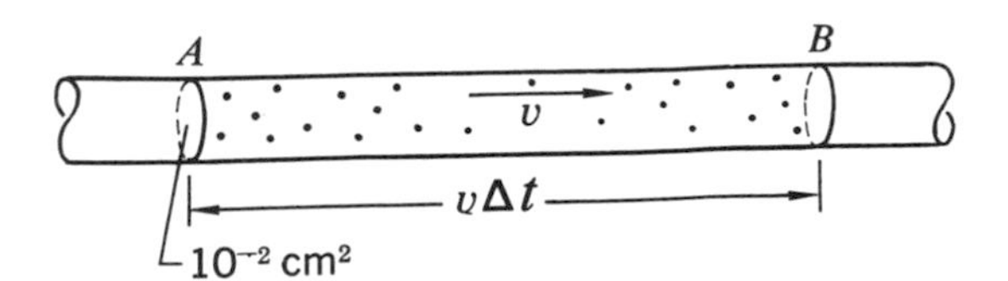

Fig. 8–14 Schematic diagram of a section of a current-bearing wire.

Equating these two expressions yields

$$v = 6 \times 10^{-2}\ \text{cm/sec} = 0.6\ \text{mm/sec}$$

The electrons are barely moving! And in terms of the above theory, it means that we have related the magnetic force exerted by the electron current to the Lorentz contraction (of the distance between uniformly moving charges) under conditions such that the ratio v/c is only about 2×10^{-12} and hence the Lorentz contraction is almost inconceivably small—only of the order of 1 part in 10^{23}. We are accustomed to thinking of magnetic forces as being quite large and important, yet we can ascribe them to the effect of these seemingly negligible relativistic contraction effects at low speeds.[1] Who says relativity is important only for velocities comparable to that of light?

Another implication of the above analysis is that any departure from electrical neutrality of a current-bearing wire, as observed in its own rest frame, must be very small indeed, or else the electric force on a moving charge outside the wire would completely swamp the magnetic force.

MORE ABOUT MAGNETIC FORCES AND LORENTZ CONTRACTIONS

For the situation considered in the last section, in which the test charge was assumed to have the same velocity as the electrons in the wire, a space-time diagram helps to provide a very direct appreciation of the way in which the test charge "sees" the wire as carrying a net positive charge. Figure 8–15 shows the world lines of the positive ions and the electrons. The positive ions are at rest in S and the electrons are moving. Their densities are the same and therefore the distances between the ions and between the electrons, measured at any time t, are equal.

[1]It is important to note that *no* contractions are involved from the standpoint of the laboratory frame, but only from the standpoint of a frame moving relative to the laboratory. The only difference between a wire carrying a current and a wire not carrying a current is the existence of a drift velocity for the electrons. The mean distance between the electrons remains unaffected as measured in the laboratory frame.

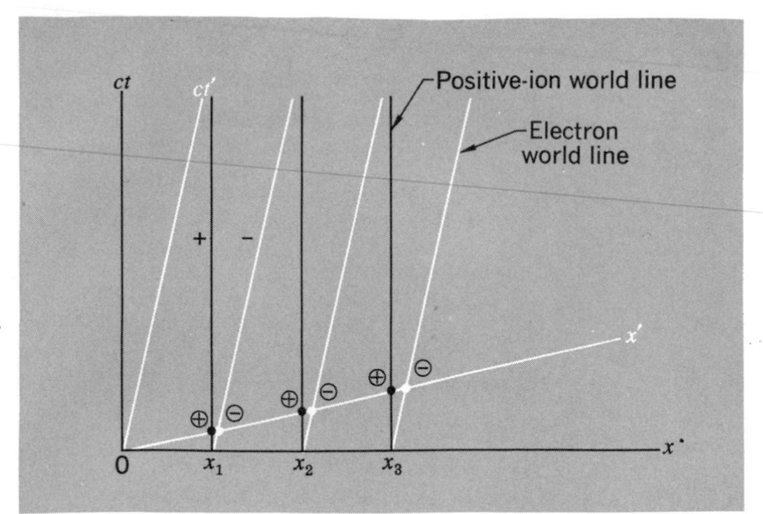

Fig. 8–15 Minkowski diagram to illustrate how a neutral current-bearing wire appears to carry a net charge density as observed in a moving reference frame.

The S' frame is moving relative to S with the same velocity as the electrons. On our space-time diagram, the ct' axis is therefore parallel to the world lines of the electrons and the angle between the x' and x axes is then the same as that between the ct' and ct axes. In S', the location of the positive ions and the electrons at $t' = 0$ is given by the intersection of their world lines with the x' axis. It is clear that in S' the distances between the positive ions are less than the distances between the electrons, and these respective distances remain the same at any time t'. Thus we again see that in S' the density of the positive ions in the wire is greater than the density of electrons, and therefore a stationary electron outside the wire will experience an electric force. As before, it is this force which, in S, is perceived as a magnetic force. With the use of the appropriate scale for distances measured along the x' axis (refer to the discussion of Minkowski diagrams in Chapter 3), the space-time diagram will of course yield the correct quantitative values for λ_+ and λ_-.

The situation that we considered above was very special, because we assumed that the test charge and the line of electrons both had the same velocity v. Let us now take the case of a test charge moving parallel to the neutral current-bearing wire at some velocity $\mathbf{u}$ which is different from the electron drift velocity $\mathbf{v}$ [Fig. 8–16(a)]. What force acts on it? Since the wire is neutral and the positive ions, being stationary, produce no magnetic field, we already have the answer; it is equal to the purely magnetic force due to the moving line of electrons, according to

Eq. (8–13):

$$F_{\text{mag}} = -q_2 \frac{uv}{c^2}\left(2k\frac{\lambda}{b}\right)$$

But let us briefly show how this result, too, can be understood in terms of the relativistic modification of the positive and negative charge densities.

As before, we shift to the frame S' in which the electron test charge is at rest and therefore sensitive only to the electric force [Fig. 8–16(b)]. In this frame the positive ions in the wire have the velocity $-\mathbf{u}$ and the electrons have a velocity $\mathbf{v}'$ given by the velocity-addition law:

$$v' = \frac{v - u}{1 - uv/c^2}$$

In S, the magnitudes of the positive and negative charge densities are both λ_0. But in S' we have

$$\lambda_+ = \gamma(u)\lambda_0$$

$$\lambda_- = -\frac{\gamma(v')}{\gamma(v)}\lambda_0$$

(Satisfy yourself, in particular, that the equation for λ_- makes sense. Imagine going into the rest frame of the electrons and then out again.) But we have the relationship

$$\gamma(v') = \gamma(u)\gamma(v)(1 - uv/c^2)$$

[refer to Eq. (7–7)]. Thus we have

$$\lambda_- = -\gamma(u)(1 - uv/c^2)\lambda_0$$

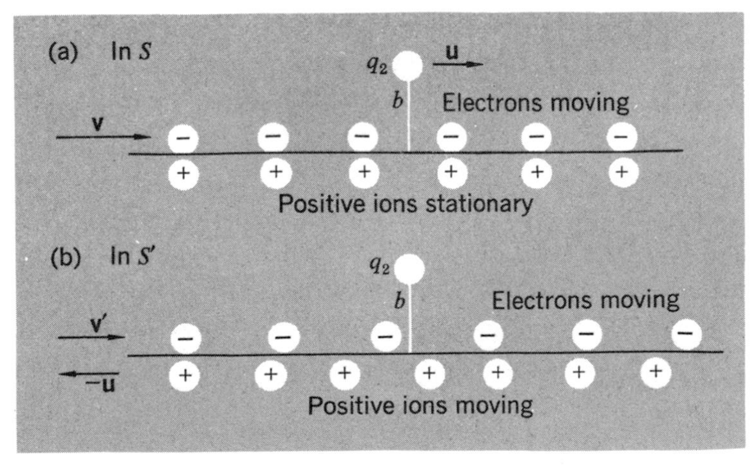

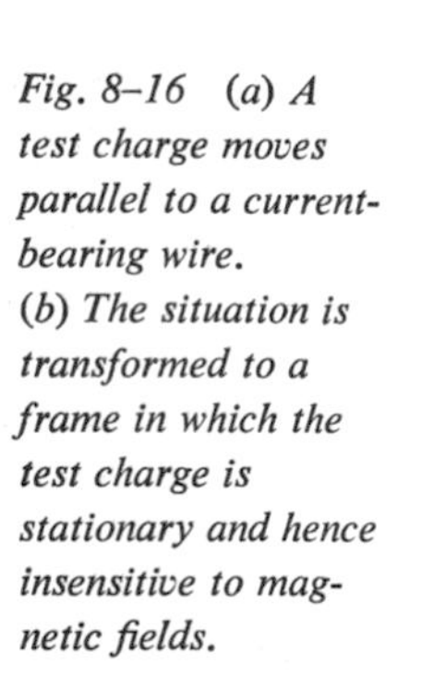

Fig. 8–16 (a) A test charge moves parallel to a current-bearing wire. (b) The situation is transformed to a frame in which the test charge is stationary and hence insensitive to magnetic fields.

The net charge density along the wire, as measured in S', is thus given by

$$\lambda' = \lambda_+ + \lambda_- = \gamma(u)\frac{uv}{c^2}\lambda_0$$

Proceeding exactly as in the previous more special case, we find that the net force on q_2, as measured in the laboratory frame S, is given by

$$F_y = \frac{uv}{c^2}\left(\frac{2k\lambda_0}{b}\right)q_2$$

If you are bothered by the absence of a negative sign in this equation, as compared to Eq. (8–13), remember that the charge density of the *moving* charges in the wire, i.e., the electrons, is equal to $-\lambda_0$, and putting $\lambda = -\lambda_0$ in Eq. (8–13) leads exactly to the result above.

THE FORCE BETWEEN CURRENT-BEARING WIRES

We are now ready for the culmination (so far as we are concerned in this book) of the relativistic analysis of the forces between charges in various states of motion. We shall consider the force between two long parallel metallic wires a distance b apart, carrying currents I_1 and I_2 (Fig. 8–17). The densities of positive and negative charges are equal in each of the wires, and therefore each wire produces no electric field. But each wire does produce a magnetic field because there is an electron flow, and this magnetic field exerts a force on the moving electrons in the other wire.

The basis of our calculation is the statement of the force exerted on a test charge q_2, traveling at speed u parallel to an infinitely long line of charge moving at speed v:

$$F_y = -\frac{uv}{c^2}\left(\frac{2k\lambda}{b}\right)q_2$$

We can describe this, if we wish, in terms of the magnetic field

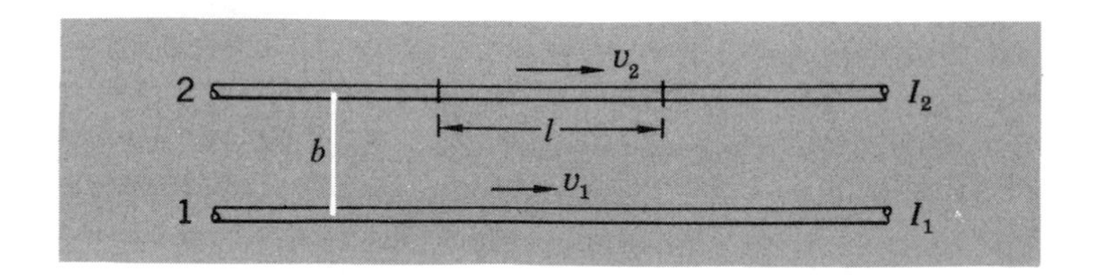

Fig. 8–17 Two parallel current-bearing wires.

at the position of q_2, as given by Eq. (8–18):

$$B_z = \frac{v}{c^2}\left(\frac{2k\lambda}{b}\right)$$

giving

$$F_y = -q_2 u B_z$$

To apply these results to the present problem we must introduce the relation between current, linear charge density, and velocity of the moving charges:

$$I = \lambda v$$

(Prove this.) Let the magnitudes of linear charge density and electron drift velocity in the two wires be λ_1, v_1 and λ_2, v_2, respectively. Then the magnetic field due to wire 1 at the position of wire 2 is given by

$$B_z = \frac{v_1}{c^2}\left(\frac{2k\lambda_1}{b}\right) = \frac{2kI_1}{c^2 b} \tag{8–19}$$

and is in the direction perpendicular to the plane containing the wires. Let us calculate the force that this field exerts on all the electrons in a length l of wire 2. The amount of moving charge (q_2) is numerically equal to $\lambda_2 l$, and its velocity (u) is in this case v_2. Hence the magnitude of the force on this length of wire is given by

$$F = \lambda_2 l v_2 \left(\frac{2kI_1}{c^2 b}\right)$$

i.e.,

$$F = \frac{2k}{c^2}\frac{I_1 I_2}{b} l = B_z I_2 l \tag{8–20}$$

This force, as we have seen, is in the plane of the wires. If the currents are in the same direction, the force is attractive. If the currents are oppositely directed, the force is repulsive.

The constant k, remember, is the constant of proportionality in Coulomb's law. We see now that it appears in the expression for the force between two current-carrying wires and can therefore be used to calculate that force when known currents are flowing. In the MKS system, as we mentioned at the beginning of the chapter, its value is about 9×10^9 newton $\cdot$ m^2/coulomb^2. Using Eq. (8–20), one can calculate from this that the force per unit length between two parallel wires, 1 cm apart, each carrying a current of 10 amp, is 0.002 newton/m. If there were no positive ions to neutralize the electric field of the electrons, the electric force between the wires would be a factor of c^2/v^2 greater than this, or about 5×10^{20} newton/m!

UNITS AND SYSTEMS OF ELECTRIC AND MAGNETIC MEASUREMENT

To wrap up this discussion, as it were, let us return to the quantitative statements of the electric and magnetic force laws with which we began the chapter. The subject of basic units, etc., may often seem rather dull, but the unification of electricity and magnetism through relativity provides a special interest in this case.

Equation (8–1) states the basic law of force between electric charges (Coulomb's law). The last equation [Eq. (8–20)] states a law of force between parallel currents. The CGS unit of charge was defined by setting the constant k equal to unity in the Coulomb law. But what if we let Eq. (8–20) define our unit of charge? We can do this by asserting that two parallel wires, placed a unit distance apart and each carrying a current of unit strength, repel or attract one another with exactly 2 units of force per unit length. Why 2 units of force and not 1? To correspond with the factor 2 on the right side of Eq. (8–20). Historically, this was in effect done to define an absolute unit of current, and the units of mechanical measurement were CGS units.[1] Thus the definition in effect read as follows:

> If two long parallel wires, placed 1 cm apart, each carry a unit of current, the force exerted on each wire by the other is 2 dynes per cm of its length.

The unit of current thus defined is actually 10 amp in our practical units of measurement—i.e., a rate of flow of charge of exactly 10 coulombs/sec. Let us now restate the above definition, using MKS units:

> If two long parallel wires, placed 10^{-2} m apart, each carry current at the rate of 10 coulombs/sec, the force exerted on each wire by the other is 2×10^{-5} newton/10^{-2} m of its length.

Once we have done this, we have completely determined the value of the constant k in what we familiarly call the MKS system.

[1]We say "in effect" because, for practical reasons, the measurements were made on the force between current-bearing coils rather than between long straight wires, and a formula appropriate to this other geometry was used.

For, by substitution in Eq. (8–20), we have

$$2 \times 10^{-5} = \frac{2k}{c^2} \frac{10 \times 10}{10^{-2}} \times 10^{-2}$$

Therefore,

$$k = 10^{-7}c^2 \text{ newton} \cdot \text{m}^2/\text{coulomb}^2$$

But

$$c = 2.9979 \times 10^8 \text{ m/sec} \qquad \text{(see Table 1–2)}$$

Therefore,

$$k \text{ (MKS)} \approx 8.987 \times 10^9 \text{ newtons} \cdot \text{m}^2/\text{coulomb}^2 \qquad (\text{i.e. } \approx 9 \times 10^9)$$

Putting this result another way, we can see how, given an experimental knowledge of the electric force between stationary charges and the magnetic force between currents, we are in a position to evaluate the magnitude of c itself. That is just what Clerk Maxwell did in the 19th century, when he came to recognize that the laws of electricity and magnetism, as established in the laboratory, implied the possibility of electromagnetic waves traveling at the speed c through space.

And now let us add a few final remarks about the electric and magnetic fields themselves. We have seen in some detail the confirmation of Einstein's conviction that a magnetic force is none other than an electric force from a different point of view. Why not, then, apply this same logic to the fields as such? That is, define a magnetic field as being dimensionally the same as an electric field. It is not difficult to arrange this. Let us see how it can be done.

We shall return to the basic calculations of the force exerted by one moving charge on another. If charge q_1 moves with velocity $\mathbf{v}$, the net force that it exerts on charge q_2, moving with velocity $\mathbf{u}$, is given, according to the general force law [Eq. (8–4)] by

$$\mathbf{F}_{12} = q_2(\boldsymbol{\mathcal{E}} + \mathbf{u} \times \mathbf{B}) \qquad \text{(MKS system)}$$

where, according to Eq. (8–10), we have

$$\mathbf{B} = \frac{1}{c^2}(\mathbf{v} \times \boldsymbol{\mathcal{E}}) \qquad \text{(MKS system)}$$

Now our introduction of the fields $\boldsymbol{\mathcal{E}}$ and $\mathbf{B}$ is, in a way, just a device to allow us to describe the force on an arbitrarily moving test charge in convenient terms. If we choose to assert that electric and magnetic fields are really quantities of the same kind, we can arrive at the same value of the physically observable quantity—the force **F**—by simply transferring a factor $1/c$ from the definition of **B** to the statement of the force law. And this is precisely how the equations are written in what is traditionally called the CGS system—although, as mentioned at the beginning of the chapter, the essential point is not the choice of units as such but the use of Coulomb's law, rather than the force between currents, to define a unit of charge. Thus in the CGS system the two equations above are replaced by the following:

$$\left.\begin{aligned} &\mathbf{F}_{12} = q_2\left(\boldsymbol{\mathcal{E}} + \frac{\mathbf{u}}{c} \times \mathbf{B}\right) \\ &\text{where} \\ &\mathbf{B} = \frac{\mathbf{v}}{c} \times \boldsymbol{\mathcal{E}} \end{aligned}\right\} \text{(CGS system)}$$

The magnetic field is thus defined, in effect, as some fraction ($\approx v/c$) of the electric field. We see, then, how it comes about that in the magnetic force law,

$$\mathbf{F}_{\text{mag}} = \text{const.}\ (q_2\mathbf{u} \times \mathbf{B})$$

the value of the constant should be unity according to a system (MKS) in which unit charge is defined in terms of the force between currents, but equal to $1/c$ according to a system (CGS) in which the unit charge is defined in terms of Coulomb's law.

It is apparent that the CGS system has much to commend it from the standpoint of the lack of any fundamental distinction between electric and magnetic forces. But whichever system one uses, the results that we have developed make it possible to formulate a set of transformation equations between components of $\boldsymbol{\mathcal{E}}$ or **B** in one frame and related components of both $\boldsymbol{\mathcal{E}}$ and **B** in another frame. They will be a set of linear equations, essentially similar to the original Lorentz transformations that mingle distance and time, or the dynamical transformations that mingle momentum and energy. You can easily develop them if you choose. But their general application would be outside the scope of this book.

PROBLEMS

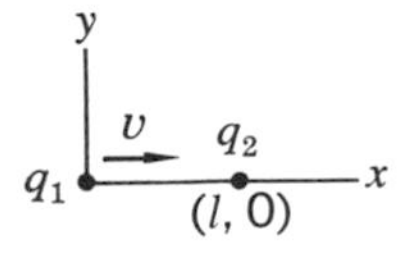

8–1 Consider the situation in the figure. Show, by transforming into the rest frame of q_1 and then back into the laboratory frame, that the force on the test charge q_2 is the same whether the source charge q_1 is moving toward q_2 or receding from it.

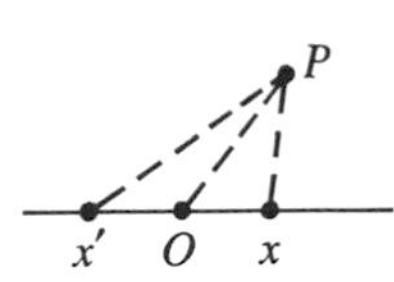

8–2 (a) A charge q moves with constant velocity **v**, and at time t is at the point x (see the figure). The point x' is the position of q at an earlier time t', such that a light pulse emitted from x' at time t' arrives at P at time t (i.e., x' is the retarded position of q with respect to an observation at P at time t). Show that a signal emitted from q as it passes point 0 arrives at P after time t.

(b) Suppose q moves with arbitrary velocity (over an arbitrary path), but with $v < c$. Show that there is one and only one retarded position with respect to an observation at point P at time t.

8–3 If a charge is at the origin at a certain instant, moving with velocity v along x, the electric fields at $\pm x$ at that instant are equal. The text (p. 242) comments on the remarkable fact of this fore-and-aft symmetry. Intuitively, one might have expected the field to be weakened at points from which the charge is receding and strengthened at points which it is approaching. The basic soundness of this intuition becomes apparent if one takes account of the time needed to propagate the field. To consider this problem, suppose that a charge of velocity v ($= \beta c$) passes through the origin at time $t = 0$. The field at $x = a$ at $t = 0$ is associated with a previous (retarded) position of the charge, $x = -b$, such that $b/v = (a + b)/c$. Similarly, the field at $x = -a$ is associated with a different retarded position, $x = -d$. Express the electric field in terms of the distance between the field point and the appropriate *retarded* position of the charge in each case. Deduce that, if we had considered a situation for which the retarded separations had been *equal*, the field at the point ahead of the charge would exceed that at the point behind the charge by a factor equal to $(1 + \beta)^2/(1 - \beta)^2$.

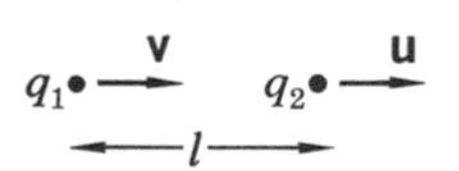

8–4 Consider the situation in the figure, in which a source charge moving with velocity **v** acts on a test charge moving in the same direction, and along the line joining them, with speed u (see the figure). What is the force on the test charge q_2 due to the source charge q_1? Why is this case not very instructive for the analysis of magnetic forces?

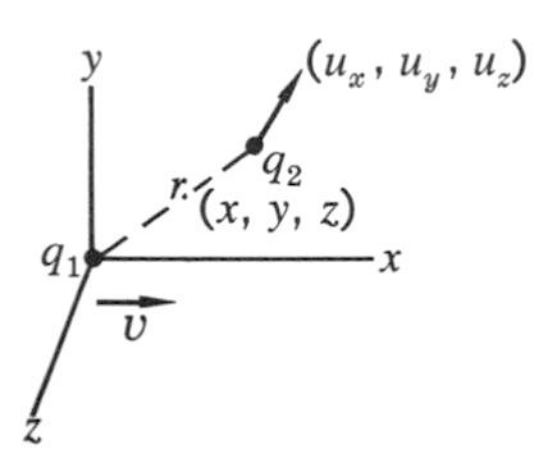

8–5 A source charge q_1 situated at the origin moves in the x direction with velocity **v** (see the figure). What is the force on a test particle q_2

situated at an arbitrary position **r**, moving with an arbitrary velocity **u**? Show that $\mathbf{B} = \mathbf{v} \times \mathcal{E}/c^2$ for this situation.[1]

8–6 Two inertial frames S and S' move relative to one another with speed w. Derive the transformation equations relating the electric and magnetic fields in the two frames of a point charge q moving with constant velocity in the x direction.

8–7 An infinitely long line charge of density λ is situated on the x axis of a certain coordinate system. This line charge moves in the x direction with constant velocity **v**. A test charge q moving with arbitrary velocity is located at the point ($x = 0$, $y = b$, $z = 0$). Find the force on the test charge and use it to show that the electric and magnetic fields at $(0, b, 0)$ are related by the equation $\mathbf{B} = \mathbf{v} \times \mathcal{E}/c^2$.

8–8 Two parallel copper wires each carry a current of 1 amp.

(a) Assuming there is one free electron for every Cu atom, what is the drift velocity of these electrons if the cross-sectional area is 0.01 mm^2? 1 mm^2? 1 cm^2?

(b) What is the force per unit length experienced by each wire if they are separated by a distance of 1 m? 1 cm?

8–9 It is pointed out in the text that when the free electrons in a neutral wire are given a drift velocity, so that a current flows, the mean distance between the electrons as observed in the laboratory remains unchanged. How do you reconcile this with the fact that all objects are subject to a Lorentz contraction when they go from a state of rest to a state of motion in a given frame?

8–10 From the geometry of the Minkowski diagram of a current-carrying wire (Fig. 8–15), calculate the ratio of positive to negative charge densities as observed in the rest frame of the electrons. [All you really need is the knowledge that the angle between the x and x' axes is $\tan^{-1}(v/c)$.]

8–11 Two inertial frames, S and S', move with speed v with respect to one another. Along the x axis of S lies an infinitely long wire which is composed of stationary positive charges, and negative charges moving in the x direction with speed v. Thus a current flows through the wire, although the net charge density in S is everywhere zero. What is the net charge density in S'? Does this result imply that total charge is not conserved in a Lorentz transformation? Explain. [See D. L. Webster, *Am. J. Phys.*, **29**, 841 (1961), for a discussion and explanation of this phenomenon.]

[1]In this and other problems in which the magnetic *field* **B** is explicitly introduced, the use of the MKS system is assumed.

Epilogue

WE SHALL END this account of special relativity—its theoretical bases, and some of its applications—by taking a brief look at the historical origins of the theory. It makes a fascinating story, many aspects of which have been recounted by Holton.[1]

Although, as we have said, Einstein has a clear right to be called the father of special relativity, there were others who could have claimed to have some part in its parentage. The man who probably came closest to anticipating Einstein was the great French theoretician H. Poincaré. He could not quite bring himself to abandon the luminiferous ether, but he firmly believed in the relativity principle and in the fundamental impossibility of detecting absolute motion by means of optical observations. He asserted this in 1899, and expanded on his ideas in an impressive paper, published in 1904, entitled "The present state and the future of mathematical physics."[2] Almost half of this paper is devoted to essentially relativistic topics—the Lorentz contraction, the increase of mass with velocity, the status of c as a limiting velocity in dynamics. He even considers the problem of comparing clocks by an exchange of light signals. It remains true, however, that his discussion is rooted in Lorentz's analysis of electromagnetic phenomena, using the notions of a real contraction of lengths and of a so-called "local time" [defined formally by equations (3–14)]. Poincaré did express his belief that the Lorentz theory was not the last word, but his remarks reveal that he was not himself ready with a superior theory.

As for Lorentz—another of the truly great theoretical physicists (he was a Nobel prize winner in 1902)—the development of his transformations in 1904 never called in question the

[1]G. Holton, *Am. J. Phys.*, **28,** 627–636 (1960).

[2]H. Poincaré, *Bull. Sci. Math.*, **28,** 302–323 (1904).

existence of the special reference frame of the ether, and 15 years later he still attached some value to the idea of absolute space. Another curious sidelight on this history is that the Lorentz transformations had, in essence, been discovered in 1887 by W. Voigt, who in that year published a theoretical paper about the Doppler effect (which can be regarded as the problem of observing a wave motion from different inertial frames).

One last item of pre-Einsteinian history is that the need for some kind of correction factor on time measured in a moving system was recognized by Sir Joseph Larmor in 1900 as a concomitant of the FitzGerald-Lorentz contraction, if one demanded invariance for the equations describing the properties of the ether.

What, then, was Einstein's uniquely important contribution? It was, as he himself recalled,[1] his insight into the problem of measuring time, and it was the result of nearly 10 years of thought and speculation, beginning at age 16! From this stemmed everything else, including the Lorentz transformations and a new dynamics in which mass varied with velocity. It is all spelled out in his paper of 1905, which by almost universal agreement identifies Einstein as the true creator of relativity theory.[2] His paper is a study in itself, not only for the richness of its content, but also for its silences. There is not a single reference to the scientific research literature. It appears quite certain that Einstein was unaware of Lorentz's paper of the year earlier; he rederived the Lorentz transformations from scratch. But the really tantalizing question is this: On what kind of a basis did Einstein in fact build his theory? When asked about this many years later, he said that he had been influenced by certain well-known optical observations (the aberration of starlight and the propagation of light in rapidly flowing water) but did not recall having had any knowledge, prior to 1905, of the Michelson-Morley experiment.[3]

[1]See his autobiographical memoir in *Albert Einstein: Philosopher-Scientist* (P. A. Schilpp, ed.), Harper Torchbooks, New York, 1959.

[2]It is one of the minor mysteries of scientific history that Sir Edmund Whittaker, in the remarks quoted at the head of Chapter 3, should have been so unready to give Einstein his due. (He had no such reservations when it came to the general relativity theory.) Whittaker was a generally scrupulous and impeccable scholar, and his *History of the Theories of Aether and Electricity* is a most fascinating and richly documented account of the development of physical thought—strongly recommended as background reading.

[3]See R. S. Shankland, *Am. J. Phys.*, **31,** 47 (1963). But see also Shankland's later article about the Michelson-Morley experiment in *Am. J. Phys.*, **32,** 16 (1964). This quotes a letter written by Einstein in 1952 in which he *did* profess to have known about the Michelson-Morley experiment before 1905,

Yet this experiment was a classic; it had been performed 18 years earlier, and it was carried out in direct response to Maxwell's speculations on the possibility of detecting the earth's motion through space. One would think that by 1905 the results of the Michelson-Morley experiment, if not its details, must have been familiar to all physicists interested in such questions. Even if Einstein did not consciously recall the experiment, it is hard to believe that his thinking was not in some way influenced by it. When all this has been said, however, it must be admitted that Einstein had an extraordinary insight. He seemed to know the answers without benefit of anything more than a hint or two from nature; the rest was mere confirmation. This, indeed, was the characteristic mark of his genius. He proceeded in almost the same way (also in 1905) in setting forth the photon hypothesis and the photoelectric equation on the basis of scanty and inconclusive data. It was for this latter work, incidentally, and not (at least nominally) as the creator of relativity that he was awarded the Nobel prize for physics in 1921.[1] But relativity, taking the special and the general theories together, will undoubtedly continue to be regarded as the chief monument to Einstein's life in physics. (He published the general theory in 1916.) It is, in the words of C. Møller,[2] "one of the most beautiful chapters in the history of science, which for the main part was written by a single man . . . ."

---

as a result of reading Lorentz's papers. Apparently his recollection of these matters, more than five decades after the events, was not entirely consistent. But he claimed on both occasions that other pieces of evidence played the largest part in shaping his thinking.

[1]Alfred Nobel's will stipulated that the prizes were to be for discoveries conferring benefit on mankind. It was apparently doubted whether a pure theory met this requirement, and the citation for the award read "For his contributions to mathematical physics, and especially for his discovery of the law of the photoelectric effect."

[2]C. Møller, in the preface to his book, *The Theory of Relativity*, Oxford Univ. Press, New York, 1952.

A short bibliography

The literature of relativity is enormous—probably more, in relation to its basic content, than for any other area of physics. Below is a brief and inevitably somewhat arbitrary list of books that may be of interest as background and parallel reading for this book and for various topics that have been discussed in it.

Bergmann, P. G., *Introduction to the Theory of Relativity*, Prentice-Hall, Englewood Cliffs, N.J., 1942.

A standard text, devoted about equally to the special and the general theories, with considerable emphasis on tensor calculus.

Bohm, D., *The Special Theory of Relativity*, Benjamin, New York, 1965.

A general discussion at about the level of the present book, enriched by Bohm's perennial concern with fundamentals.

Born, M., *Einstein's Theory of Relativity*, Dover, New York, 1962.

A splendid and thorough presentation at a fairly elementary but far from trivial level, by one of the greatest expositors of physics in the 20th century—and a Nobel prize winner into the bargain (for his fundamental contributions to quantum theory).

Eddington, Sir A. S., *Space, Time and Gravitation*, Harper Torchbook, Harper & Row, New York, 1959.

Mostly about the general theory.

Einstein, A. (translated by R. W. Lawson), *Relativity, the Special and the General Theory*, Crown, New York, 1961.

A clear and simple presentation of the basic ideas, by their originator.

Einstein, A., and Infeld, L., *The Evolution of Physics*, Simon and Schuster, New York, 1961.

A completely nonmathematical account of the growth of the most basic concepts and results in physics, with a strong emphasis on relativity.

Einstein, A., and others (translated by W. Perrett and G. B. Jeffery), *The Principle of Relativity*, Dover, New York, 1958.
A collection of the historic papers, mainly by Einstein, Lorentz, and Minkowski, through which the theory came into existence.

Ford, K. W., *The World of Elementary Particles*, Ginn, Boston, 1963.
A lively and imaginative presentation of the principles and phenomena of high-energy physics.

Frisch, D. H., and Thorndike, A. M., *Elementary Particles*, Van Nostrand, Princeton, N.J., 1964.
Very much like Ford's book in its general coverage and liveliness, but written more from the experimentalist's standpoint.

Jaffé, B., *Michelson and the Speed of Light*, Anchor Books, Doubleday, New York, 1960.
Mainly biographical, with interesting sidelights on some of the key experiments performed by Michelson.

Kacser, C., *Introduction to the Special Theory of Relativity*, Prentice-Hall, Englewood Cliffs, N.J., 1967.
A presentation at much the same level as the present book.

Katz, R., *An Introduction to the Special Theory of Relativity*, Van Nostrand, Princeton, N.J., 1964.
A good brief survey.

Michelson, A. A., *Studies in Optics*, Univ. Chicago Press, Chicago, 1927.
Michelson's own modest account of his superb investigations.

Ney, E. P., *Electromagnetism and Relativity*, Harper & Row, New York, 1962.
A brief account of the main principles, including some general relativity.

Reichenbach, H., *The Philosophy of Space and Time*, Dover, New York, 1957.
An extended discussion of the description of nature through the geometry of space-time.

Rindler, W., *Special Relativity*, Wiley (Interscience), New York, 2nd ed., 1966.
A very compact and rather mathematically oriented presentation.

Rosser, W. G. V., *An Introduction to the Theory of Relativity*, Butterworth, London, 1964.
A very thorough discussion, higher in level than any other of

the books listed here, and with a strong emphasis on the connection between relativity and electromagnetism.

———., *Introductory Relativity*, Butterworth, London, 1967.
A shortened and somewhat simplified version of the preceding reference—still a little above the level of the present book.

Schilpp, P. A. (ed.), *Albert Einstein: Philosopher-Scientist* (2 vols.), Harper & Row, New York, 1959.
A fascinating collection of nonmathematical articles, mainly by outstanding physicists (including Niels Bohr, Max Born, and Louis de Broglie) relating to Einstein's contributions to physics, and prefaced by a lengthy autobiographical memoir by Einstein.

Sherwin, C. W., *Basic Concepts of Physics*, Holt, New York, 1961.
Chapters 4 and 5 of this book are a refreshingly original presentation of the ideas of relativity and electromagnetism.

Smith, J. H., *Introduction to Special Relativity*, Benjamin, New York, 1965.
A very clear-minded discussion of principles and phenomena.

Taylor, E. F., and Wheeler, J. A., *Spacetime Physics*, Freeman, San Francisco, 1966.
A stimulating and highly readable presentation of relativistic ideas, juxtaposing formality and informality in an intriguing way.

Whittaker, Sir E. T., *History of the Theories of Aether and Electricity* (2 vols.), Harper Torchbook, Harper & Row, New York, 1960.
Traces in minute and fascinating detail the historical development of optics and electromagnetism; quite technical in its citations of mathematical theory, but with copious reference to experiments. Notable, as mentioned in the Epilogue, for the scant recognition it gives to Einstein as far as special relativity is concerned.

Whittaker, Sir E. T., *From Euclid to Eddington*, Dover, New York, 1958.
A kind of thumbnail sketch of the evolution of physical thought, with the concepts of space and time strongly represented.

The American Journal of Physics, journal of the American Association of Physics Teachers (AAPT), is a perennial source of papers and letters discussing particular points, paradoxes, etc., in relativity theory. The AAPT has also produced a booklet entitled *Selected Reprints in Special Relativity Theory* (published by the American Institute of Physics, New York, 1963) which contains in addition a categorized and annotated bibliography by G. Holton. (The reader who is interested in more specialized or advanced treatments will find numerous suggestions in Holton's list.)

Answers to problems

CHAPTER 1

1–1 6.3 °K.

1–2 (a) 5.76×10^{-3} newton; (b) about 1 newton.

1–3 (a) 6×10^{8} newtons (gravitational force = 3.6×10^{22} newtons); (b) about 2×10^{-7} m.

1–4 About 0.5 cm.

1–5 Reasonable estimates for the power to run a washing machine lead one to the result that the solar mass loss is occurring at the rate of a few million tons per second, which is about right.

1–6 Of the order of a gram, perhaps.

1–7 10^{-14} (order of magnitude).

1–9 About $50,000.

1–10 4.7×10^{13} m.

1–11 (a) 3.3×10^{-11}; (b) 6.4×10^{-4}; (c) about 10^{11} °K.

1–14 0.88″ of arc.

1–15 (a) $c[n(n+2)]^{1/2}/(n+1)$; (b) $m_0c[n(n+2)]^{1/2}$.

1–16 (a) 2000 volts; (b) $0.09c$.

1–17 (a) About 1.5; (b) about 0.75.

1–18 $c(1 + m_0c^2/h\nu)^{-1}$.

1–20 (a) 21.3 MeV; (b) 1.2×10^{-3} newtons.

1–21 (a) 10^{-16}, 10^{-12}, 10^{-9}, 10^{-2}, 0.65; (b) about $0.1c$.

1–23 Obtain the equations for v_x and v_y, taking account of the variation of mass with energy, and solve for x and y separately as functions of t.

1–24 $A = \frac{3}{10}$, $B = \frac{9}{20}$.

1–25 1.82×10^{25} sec^{-1}. No, speed at equator is greater than c.

1–26 (a) 0.0025%; (b) 35%.

CHAPTER 2

2–1 (a) 24°; (b) 61°; (c) 0.1 light-year (as compared to about 4 light-years for the actual nearest star, α Centauri).

2–3 (b) 1750 rpm.

2–4 3.3×10^{-3} Å.

2–5 (b) About 30 sec; about 16 min.

2–6 About 200 fringes.

2–7 3.5 km/sec.

2–9 (a) In most orientations $\cos 2\theta \neq 0$; (b) 4; 0°, 90°, 180°, 270°; (c) 9.5 m/sec.

CHAPTER 3

3–2 (a) 2×10^{-4} fringe; (b) 0.08 fringe.

3–3 (b) Yes.

3–5 $x = 93$ m, $t = 2.5 \times 10^{-7}$ sec.

3–6 (a) $-c/2$; (b) $\sqrt{3}\, x_0/c$.

3–7 (a) 1.25×10^{-7} sec; (b) 2.25×10^{-7} sec.

3–8 (c) (1) $x' = 0.58$, $ct' = 0.58$; (2) $x = 1.73$, $ct = 1.73$; (3) $x = 2.31$, $ct = 1.15$; (4) $x' = -1.15$, $ct' = 2.31$.

CHAPTER 4

4–1 1.34×10^9 m.

4–2 5.77×10^{-6} sec.

4–3 (a) ± 0.3 m; (b) 0.33×10^{-8} sec; (c) 0.27 m, -0.34 m.

4–4 (a) 10^5 years; (b) 5 min.

4–5 (a) l_0/c; (b) $[(1 - \beta)/(1 + \beta)]^{1/2} l_0/c$; $l_0/\gamma v$.

4–6 (a) 200 m; (b) 6.67×10^{-7} sec; (c) 3.33×10^{-7} sec.

4–7 (a) 4.67×10^{-8} sec; (b) 6.7 m.

4–8 $l[(1 - \beta)/(1 + \beta)]^{1/2}$; $(1 - \beta)\gamma l/c$.

4–9 (a) 2×10^7 m/sec; (b) 1 o'clock plus 4.99×10^{-6} sec.

4–10 $8.63E_0$.

4–11 About 0.9×10^{-8} sec.

4–12 (a) 12:50 P.M.; (b) 7.2×10^{11} m; (c) 1:30 P.M.; (d) 4:30 P.M.

4–13 (a) $0.9950c$ to $0.9954c$;
(b) $R(\theta) \approx R_0\{1 + [(1 - \cos\theta)/\gamma_0] \ln R_0\}$, where $R(\theta)$ is the count-rate ratio at θ and R_0 is the ratio at vertical incidence (for mesons having $\gamma = \gamma_0$). This result comes from using the relation $\gamma(\theta) = 1 + (\gamma_0 - 1) \sec\theta$.

4–14 Acceleration of two rockets ceases at different times in final rest frame.

4–15 (a) $(1 - \beta\cos\theta)\gamma l/c$; (b) $(1 - \beta\cos\theta)\gamma l$.

4–16 (a) $c\sqrt{7}/4$; (b) $3\sqrt{2}\, l_0/5$.

4–17 It fits through.

4–18 For event 1: (a) yes, (b) no. For event 2: (a) no, (b) yes; $v = 2.85 \times 10^8$ m/sec, if along direction of displacement between two events in S.

4–19 Causality not violated.

CHAPTER 5

5–1 $0.7c$; $0.94c$.
5–2 $0.991c$; $0.213c$.
5–5 $1 - \exp[-n\pi(R_1 + R_2)^2 vt]$;
$1 - \exp[-n\pi(R_1 + R_2)^2(u + v)(1 + uv/c^2)^{-1}(1 - v^2/c^2)^{1/2}t]$
5–6 (a) $c/3$; (b) 20 m.
5–8 (a) $\tan^{-1}[(1 - v^2/c^2)^{1/2}/(1 - \sqrt{2}v/c)]$;
(b) $\tan^{-1}[(1 - v^2/c^2)^{1/2}/(1 - \sqrt{2}v/u)]$;
(c) $\tan^{-1}[(1 - v^2/c^2)^{-1/2}]$.
5–9 (a) $\nu_0[(1 - \beta)/(1 + \beta)]^{1/2}$; (b) $\nu_0(1 - \beta)/(1 + \beta)$.
5–10 (b) $t_0(1 - \beta)/(1 + \beta)$; (c) $t_0[(1 - \beta)/(1 + \beta)]^{1/2}$.
5–11 About 50 days.
5–12 (a) $\frac{1}{11}$; (b) $11c/61$.
5–13 (a) 1.6×10^{10} light-years; no. (b) 1.7×10^{10} light-years; no.
5–14 About 1.5×10^{10} years.
5–15 (a) 3600 Å; (b) 3×10^6 years.
5–16 (a) 22 km/sec, approaching; (b) about 0.01°;
(c) 3600 km/sec; yes.
5–17 (a) $\nu_0[(1 - \beta)/(1 + \beta)]^{1/2}$; (b) $\nu_0(1 - \beta)$; (d) 2×10^{-15}.
5–18 $x = [(1 + g^2t^2/c^2)^{1/2} - 1]c^2/g$ ($g = 9.8\ \text{m/sec}^2$);
$t = \frac{20}{3}$ months, approx.
5–20 (a) 533; (b) 267; (c) A gets 1333, B gets 1067; (d) A is younger by 2 years, 8 months.
5–21 About 2×10^{-4} sec.
5–22 (a) $\beta = 0.99975$ to 0.99997; (b) (0.9 to 2.7) $\times 10^{-14}$ sec;
(c) (2.7 to 8.1) $\times 10^{-6}$ m; (d) 0.2 μsec, 0.1 μsec.

CHAPTER 6

6–1 300 MeV; 224 MeV; mass = 200 MeV; $\tan^{-1} 0.5$; $v = 0.745c$.
6–2 (a) 5.66 GeV; (b) 4.3 amu; (c) $0.187c$.
6–3 $\sqrt{17}\, m_0$.
6–4 (a) $c/(1 + m_0c^2/E)$; (b) $4m_0/\sqrt{3}$; $c/2$.
6–5 (a) $\tan^{-1}(\frac{9}{16})$; $0.836c$; (b) 4.75.
6–6 (a) $\beta = 10^{-5}$; *blue* shift of magnitude $\beta^2/2$ because of aberration; (b) *red* shift of magnitude $\beta^2/2$; (c) *no* frequency shift.
6–8 $M_0[\gamma - 1 + (\gamma^2 - 1)^{1/2}]/2$.
6–9 (a) 33 watts; (b) 3.3 m/sec; (c) 66β watts, 0.7μ watt;
(d) (i) An ever-increasing volume of space is being filled with radiation; (ii) kinetic energy is being continually taken up by the laser through its recoil.
6–11 (a) 1.131 GeV, 4 MeV; (b) about 14°.
6–12 (a) 2 GeV, $\frac{2}{3}$ GeV; (b) 2 GeV forward, $\frac{2}{3}$ GeV backward;
(c) $\frac{2}{3}$ GeV, 2 GeV; roles interchanged with respect to (a).
6–13 (a) $c/\sqrt{3}$; (b) 1.2 MeV.
6–15 (a) 84.0°; (b) 26.1°.
6–16 $c/[1 + (m_0c^2/Q)^2]^{1/2}$.
6–17 0.25 MeV (ignoring Compton recoils from nuclei).
6–18 (a) $2.5 \times 10^{20}\ \text{sec}^{-1}$; the energy is typical of nuclear gamma rays; (b) $m_0c^2\{[1 + (2h\nu/m_0c^2)^2]^{1/2} - 1\}$.

6–19 (a) 328 MeV; (b) $\theta = 104.5°$; $Q = 228$ MeV.
6–20 850 MeV.

CHAPTER 7

7–1 K had 396 MeV kinetic energy; π has 616 MeV kinetic energy.
7–2 2.04 MeV ($=4m_0c^2$).
7–3 17.6 GeV.
7–6 $K = [(K^* + M_0c^2)/(1 - V^2/c^2)^{1/2}] - M_0c^2$, where $M_0 =$ sum of rest masses.
7–7 (a) $E' = E \cos \alpha$; (b) $E_{absorber} = E_{source} \cos \alpha = \gamma E_{source}$.
7–8 $x = [(1 + g^2t^2/c^2)^{1/2} - 1]c^2/g$ ($g = 9.8$ m/sec^2); $t = \frac{20}{3}$ months, approx.
7–9 (a) 23 mc/sec; (b) 9.7 MeV, 4.3×10^7 m/sec; (c) 49 rev, 0.34 μsec; (d) the orbital frequency of the ions decreases because of the increase of mass at the higher energies, and the resonance condition is lost.
7–10 (a) The (initial) momentum in the direction of the incident K^- meson is zero; (b) Sigma Minus.

CHAPTER 8

8–4 $F_x = kq_1q_2/\gamma^2x^2$; $F_y = 0$; $\mathbf{F}_{magnetic} = 0$.
8–5 $F_x = \gamma kq_1q_2(x + vu_yy/c^2 + vu_zz/c^2)/(\gamma^2x^2 + y^2 + z^2)^{3/2}$
$F_y = \gamma kq_1q_2y(1 - vu_x/c^2)/(\gamma^2x^2 + y^2 + z^2)^{3/2}$
$F_z = \gamma kq_1q_2z(1 - vu_x/c^2)/(\gamma^2x^2 + y^2 + z^2)^{3/2}$
8–6 Using MKS system:
$\mathcal{E}_x' = \mathcal{E}_x$ $\quad B_x' = B_x$
$\mathcal{E}_y' = \gamma(\mathcal{E}_y - wB_z)$ $\quad B_y' = \gamma(B_y + w\mathcal{E}_z/c^2)$
$\mathcal{E}_z' = \gamma(\mathcal{E}_z + wB_y)$ $\quad B_z' = \gamma(B_z - w\mathcal{E}_y/c^2)$
8–7 $F_x = 2k(q_2\lambda/b)(vu_y/c^2)$; $F_y = 2k(q_2\lambda/b)(1 - vu_x/c^2)$; $F_z = 0$.
8–8 (a) 7.4×10^{-3} m/sec, 7.4×10^{-5} m/sec, 7.4×10^{-7} m/sec; (b) 2×10^{-7} newton/m, 2×10^{-5} newton/m.
8–9 The proper distance between electrons must increase.
8–10 $1 - \beta^2$.
8–11 $I\beta(1 - \beta^2)^{-1/2}/c$. Total charge is conserved, but one must consider the complete circuit, not the charge density along a limited section.

Index